PERGAMON INTERNATIONAL LIBRARY
of Science, Technology, Engineering and Social Studies
The 1000-volume original paperback library in aid of education, industrial training and the enjoyment of leisure
Publisher: Robert Maxwell, M.C.

CRYSTALLOGRAPHY
An Introduction for Earth Science (and Other Solid State) Students

THE PERGAMON TEXTBOOK
INSPECTION COPY SERVICE

An inspection copy of any book published in the Pergamon International Library will gladly be sent to academic staff without obligation for their consideration for course adoption or recommendation. Copies may be retained for a period of 60 days from receipt and returned if not suitable. When a particular title is adopted or recommended for adoption for class use and the recommendation results in a sale of 12 or more copies, the inspection copy may be retained with our compliments. The Publishers will be pleased to receive suggestions for revised editions and new titles to be published in this important International Library.

Other Related Pergamon Titles of Interest

Books

ANDERSON & OWEN
The Structure of the British Isles, 2nd Edition

ANDERSON
The Structure of Western Europe

CONDIE
Plate Tectonics and Crustal Evolution

OWEN
The Geological Evolution of the British Isles

* RAMDOHR
The Ore Minerals and Their Intergrowths, 2nd Edition (2 vols.)

* RIDGE
Annotated Bibliographies of Mineral Deposits in Africa, Asia (exclusive of the USSR) and Australasia

ROCKS OF THE WORLD

* SAND & MUMPTON
Natural Zeolites (Occurrence, Properties, Use)

SIMPSON
Rocks and Minerals

SPRY
Metamorphic Textures

* VAN OLPHEN & FRIPIAT
Data Handbook for Clay Materials and other Non-metallic Minerals

Journals

Computers & Geosciences
Geochimica et Cosmochimica Acta
Journal of Structural Geology
Organic Geochemistry

The terms of our inspection copy service apply to all the above books except those marked with an asterisk (*). Full details of all books and journals and a free specimen copy of any Pergamon journal available on request from your nearest Pergamon office.

CRYSTALLOGRAPHY
An Introduction for Earth Science (and Other Solid State) Students

by

E. J. W. WHITTAKER

Reader in Mineralogy, Department of Geology and
Mineralogy, Oxford University

PERGAMON PRESS
OXFORD · NEW YORK · TORONTO · SYDNEY · PARIS · FRANKFURT

U.K.	Pergamon Press Ltd., Headington Hill Hall, Oxford OX3 0BW, England
U.S.A.	Pergamon Press Inc., Maxwell House, Fairview Park, Elmsford, New York 10523, U.S.A.
CANADA	Pergamon Press Canada Ltd., Suite 104, 150 Consumers Rd., Willowdale, Ontario M2J 1P9, Canada
AUSTRALIA	Pergamon Press (Aust.) Pty. Ltd., P.O. Box 544, Potts Point, N.S.W. 2011, Australia
FRANCE	Pergamon Press SARL, 24 rue des Ecoles, 75240 Paris, Cedex 05, France
FEDERAL REPUBLIC OF GERMANY	Pergamon Press GmbH, 6242 Kronberg-Taunus, Hammerweg 6, Federal Republic of Germany

Copyright © 1981 E. J. W. Whittaker

All Rights Reserved. No part of this publication may be reproduced, stored in a retrieval system or transmitted in any form or by any means: electronic, electrostatic, magnetic tape, mechanical, photocopying, recording or otherwise, without permission in writing from the publishers.

First edition 1981

British Library Cataloguing in Publication Data
Whittaker, E J W
Crystallography for mineralogists.–(Pergamon international library).
1. Crystallography   2. Mineralogy
I. Title
548′.024549   QD905.2   80–41188

ISBN 0–08–023805–X Hardcover
ISBN 0–08–023804–1 Flexicover

Preface

THE physical properties of minerals, especially those occurring as *crystals*, have fascinated natural philosophers for centuries, as first recorded by Pliny in his *Natural History* (vol. XXXVII). Crystal form became one of the most useful diagnostic properties of minerals although often quite misleading if relied upon alone without reference to other properties. Serious study of crystals and crystallography commenced in the seventeenth century with the work of Johannes Kepler (1611), Nicolas Steno (1669) and more seriously by Romé de l'Isle (1736), Haüy (1684) and many others. These studies led to the derivation of crystal lattices (Bravais, 1850), crystal symmetry, the classification into crystal systems and classes, and the derivation of space groups by Schönflies and von Federov (both in 1891). The quite independent and apparently unrelated discovery of "x-radiation" was made by Roentgen in 1895. Prior to 1912 (or much later in many universities around the world), as recorded by Max von Laue for Germany, virtually all teaching of crystallography was handled in mineralogy as a descriptive property essential as an aid to identification. The revolution in crystallography (1912) resulted from a discussion between Ewald (a graduate student at Munich) and von Laue in which the latter suggested that, if the constituent atoms of a crystal really form a lattice, irradiation of a crystal by X-rays should result in interference phenomona. The experiments using first copper sulphate and later sphalerite were successful in showing diffraction of X-rays by crystals.

Subsequent research quickly showed that analysis of the interference phenomenon could lead to a formulation of the precise structural arrangement of atoms in each crystalline compound, and thus a whole new field of research developed in crystallography. In the present book the author devotes the first part to the historically earlier and fundamental part of crystallography and the second part to the methods and basic results possible with newer techniques.

This book corresponds essentially to lectures given to undergraduate students in geology and earth sciences at Oxford University. A number of novel features are included which have been developed by the author in response to difficulties and attitudes of students who are unlikely to have a primary interest in crystallography.

The presentation is divided into two parts, chapters 1–7, comprising Part I, covers the subject-matter of a one-term course given in first year at Oxford University, while chapters 8–18 are the basis of a one-term course given in second year. Part I deals with morphological crystallography, commencing with the relation of crystal faces to "stacked bricks" or cells; repeating patterns and lattices (limited to seven P lattices, including the rhombohedral); symmetry and systems, developed by inserting symmetry into any triclinic primitive cell; rotation for faces, forms and zones, stereographic projections, axial ratios and minimum spherical trigonometry for their calculation; morphology in seven systems; derivation of symmetry and forms in all thirty-two crystal classes, with simplification of complex form and class nomenclature; properties of rotation–inversion axes.

In this book the seven hexagonal classes and five trigonal classes are grouped in hexagonal and trigonal systems respectively. However, the hexagonal four-index symbols (often called Bravais symbols) are used for all hexagonal and trigonal crystals. Brief mention of three-index symbols (Miller) referred to rhombohedral axes is made in Chapter 6.

Part II, comprising Chapters 8–18, deals with the fundamental features of X-ray crystallography, commencing with: Bragg's Law; X-ray powder diffraction; intensities of X-ray reflections, derivation of the structure factor equation with a general description of other factors; the fourteen Bravais lattices; interpretation of powder photographs, indexing; X-ray diffraction by single crystals, rotation photographs, layer lines, reciprocal lattice and graphical solution, oscillation photographs; symmetry in repeating patterns, glide planes, screw-axes and space-group determination; determination of crystal structures, illustrated by rutile, refining of element occupancy in solid solutions; electron diffraction; irregularities in crystals twinning, disorder and dislocations; morphology reconsidered. This book does not deal with the more sophisticated X-ray single crystal methods—Weissenberg and precession—for which detailed manuals are available in other texts, nor with the X-ray diffractometer methods involving the use of the Geiger-counter, proportional counter or scintillation counter for detection of the diffracted beams, which are refinements of the usual photographic recording techniques.

L. G. BERRY

Acknowledgements

THANKS are due especially to Prof. L. G. Berry and Dr. B. C. M. Butler for reading the script and making helpful suggestions; also to Mrs. A. C. Fowler for the typing, Mr. P. Jackson for preparing the X-ray photographs, Dr. J. A. Gard for providing the electron diffraction photographs, Mr. R. McAvoy for the photographic reproductions, and Miss G. Collins for drawing Figs. 1.3, 1.4 and 1.5.

Permission to reproduce or adapt illustrations from published works is gratefully acknowledged to the following:

Academic Press and Professor J. Zussman for Fig. 9.9, from *Physical Methods in Determinative Mineralogy*.

Professor I. Kostov for Figs. 7.14, 7.22, 7.43 and 7.48, from *Mineralogy* (Oliver & Boyd).

John Wiley & Sons Inc. for Figs. 7.13, 7.15, 7.17, 7.23, 7.26, 7.44 and 7.47, from *Dana's Textbook of Mineralogy* (4th Ed.) by W. E. Ford.

Hutchinson Publishing Group Ltd. for Figs. 7.4, 7.7, 7.18, 7.27, 7.34, 7.38 and 7.42, from *An Outline of Crystal Morphology* by A. C. Bishop.

International Centre for Diffraction Data for Figs. 9.10 and 9.11, from the *J.C.P.D.S. Powder Diffraction File* and *Search Manual: Minerals*.

International Union of Crystallography for Fig. 14.11, from *Symmetry Aspects of M. C. Escher's Periodic Drawings* by C. H. MacGillavry.

Longman Group Limited for Figs. 7.19, 7.40 and 18.1, from *An Introduction to Crystallography* by F. C. Phillips.

The Institute of Physics for Figs. 5.9 and 13.10 reproduced from charts PC18 and PC37P.

The Mineralogical Society of America for Figs. 7.35 and 17.15 from *American Mineralogist*.

Contents

Introduction ... xii

Part I ... 1

Chapter 1. External Form and Cell Structure of Crystals ... 3
    Conventions regarding angles ... 10
    Problems ... 11

Chapter 2. Repeating Patterns and Lattices ... 12
    Problems ... 16

Chapter 3. Crystal Symmetry ... 17
    Symmetry of unit cells in two dimensions ... 19
    Symmetry of unit cells in three dimensions ... 20
    The minimum symmetry characterising a crystal system ... 23
    Instructions for viewing stereoscopic drawings ... 27
    Problems ... 28

Chapter 4. Notation for Crystallographic Faces, Forms and Zones ... 29
    Crystallographic axes ... 29
    Miller indices ... 30
    Forms and form indices ... 30
    Symbols for zones and zone axes ... 32
    Calculations using indices and zone symbols ... 32
    Reformulation of the process of finding axial ratios ... 34
    Problems ... 35

Chapter 5. Use of the Stereographic Projection ... 36
    Well-formed crystals ... 36
    The spherical projection of a crystal ... 37
    The stereographic projection and its properties ... 38
    The stereographic net ... 42
    Projection of the lower hemisphere to the north pole ... 44
    Plotting the stereogram of a crystal ... 44
    Use of the stereogram for visualising symmetry ... 46
    Use of the stereogram of a crystal to find axial ratios ... 47
    Accurate calculations from the stereogram ... 48
    Problems ... 51

Chapter 6. Morphology of the Seven Crystal Systems ... 53
    The orthorhombic system (holosymmetric class) ... 53
    The tetragonal system (holosymmetric class) ... 56

x   Contents

	The cubic system (holosymmetric class)	58
	The monoclinic system (holosymmetric class)	66
	The triclinic system (holosymmetric class)	71
	The hexagonal system (holosymmetric class)	73
	The trigonal system (holosymmetric class)	77
	Problems	81
Chapter 7.	The Thirty-two Crystal Classes	82
	Nomenclature	84
	Representation of symmetry and derivation of forms	85
	The triclinic system	86
	The monoclinic system	87
	The orthorhombic system	89
	The tetragonal system	91
	The cubic system	96
	The hexagonal system	100
	The trigonal system	106
	Determination of crystal class	110
	Appendix on the properties of axes of rotation–inversion	115
	Problems	117

Part II		119
Chapter 8.	The Basis of X-ray Crystallography	121
	Problem	129
Chapter 9.	X-ray Powder Diffraction	130
	The Debye–Scherrer powder camera	130
	Transmission and back-reflection cameras	133
	The Gandolfi camera	134
	The Guinier camera	134
	The powder diffractometer	136
	Uses of powder diffraction	139
	Problems	144
Chapter 10.	Intensities of X-ray Reflections	145
	The atomic scattering factor	145
	Amplitude of reflection from a crystal slice	147
	Other factors affecting intensities	150
	Problems	154
Chapter 11.	The Fourteen Bravais Lattices	155
	Monoclinic system	155
	Orthorhombic system	157
	Tetragonal system	160
	Cubic system	163
	Hexagonal and trigonal systems	163
	The effect of non-primitive cells in X-ray crystallography	163
	Rhombohedral and hexagonal cells in the trigonal system	165
	Problems	166

Chapter 12.	Interpretation of Powder Photographs	168
	Powder patterns of cubic substances	169
	Powder patterns of non-cubic substances	173
	Ambiguities in indexing powder reflections	176
	Problems	176
Chapter 13.	X-ray Diffraction by Single Crystals	178
	Difficulties associated with powder photographs	178
	Rotation photographs	179
	Oscillation photographs	184
	Non-zero layer lines	186
	Problems	193
Chapter 14.	Symmetry in Repeating Patterns	194
	Space-group determination	198
	The representation of space-group symmetry	200
	The recognition of space-group symmetry	202
	Problems	205
Chapter 15.	The Determination of Crystal Structures	206
	Determination of the structure of rutile	207
	Refinement of the structure of solid solutions	213
	Problems	216
Chapter 16.	Electron Diffraction in the Electron Microscope	217
	Principles of the method	217
	Some practical details	219
	The interpretation of the diffraction pattern	220
	Problems	224
Chapter 17.	Irregularities in Crystals	225
	External form	225
	Cleavage and parting	225
	Twinned crystals	226
	Polytypes and stacking disorder	232
	Irregularities of composition	235
	Crystal defects	236
Chapter 18.	Morphology Revisited	240

Glossary of Terms Used in Crystallography	244
Answers to Problems	246
Further Reading	249
Index	251

Introduction

THE concepts of crystallography and mineralogy evolved together from the study of mineral crystals. Although some minerals are simple chemical compounds (e.g. calcite, $CaCO_3$; galena, PbS), many of the most important of them, especially the silicates, have extremely complicated chemical compositions which are very difficult to understand on the basis of a chemical analysis alone. Because of this, classifications of minerals have always relied more on the directly observable characteristics of their crystals (crystalline form, hardness, colour and optical properties) than on their composition. Indeed it is not uncommon for specimens to be assigned to the same mineral species on crystallographic grounds even though they differ from one another chemically more than they do from specimens assigned to other mineral species.

This primacy accorded by classical mineralogists to crystal morphology (i.e. the characteristics of the shapes of crystals) has been fully justified by the results of X-ray crystallography from the mid-1920s onwards. The results have revealed the detailed geometric arrangements of the atoms in minerals, and have not only made their complicated chemical compositions understandable, but have shown that the long-standing classifications based on morphology really correspond to classifications based on crystal structure at a fundamental level.

For these reasons any understanding of minerals requires an understanding of crystal structure. This in turn requires an understanding of the concepts of morphological crystallography from which it arose, and these concepts are of continuing everyday application in the practice of mineralogy, which above all demands an informed way of *just looking* at mineral specimens.

This book is written primarily for undergraduate students in geology and earth sciences, and is based on courses that have been developed to meet the needs of such students at Oxford in their first and second years. However, mineralogy is merely the first branch of solid state science to be developed, and structural crystallography is equally relevant to all students of the solid state. Because artificial materials often do not exist as crystals with well-developed external forms, chemists and physicists are generally less conscious than mineralogists of the relevance of morphological crystallography. The author believes that the lack of a background in this subject is in fact a serious hindrance to a full understanding on the part of many chemists and physicists who embark on crystallographic research, and hopes that Part I of the present book may commend itself to them as a way of making good any such deficiency.

PART I

CHAPTER 1

External Form and Cell Structure of Crystals

THE most obvious and characteristic feature of crystals is that they have flat and smooth surfaces. Each of these flat smooth surfaces is called a *face*. Different crystals have an enormous variety of shapes, but there are two features of the disposition of their faces that they have in common, and these show that the arrangement of the faces has a systematic basis.

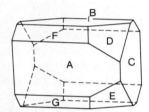

FIG. 1.1. Drawing of a crystal showing the arrangement of the faces in zones, all faces in a given zone being parallel to a particular zone axis.

1. The faces occur in sets in which all the faces in any one set are parallel to some particular direction. The crystal shown in Fig. 1.1 has the front faces lettered for reference, and the corresponding opposite faces at the back may be denoted by the same letters with primes, A', B', etc. The crystal has a very obvious set of faces ACA'C' all parallel to a vertical direction. This is the same as saying that if the faces of this set intersect one another they do so in parallel vertical *edges* of the crystal. If we look for other sets of parallel edges, then we find a set of faces parallel to the edge between A and F, namely GAFBG'A'F'B', and another to the edge between B and D, namely BDEB'D'E'. Any such set of faces on a crystal is called a *zone*, and the faces in a zone are therefore all parallel to a certain direction, which is called the *zone axis*. Some zones contain many large faces and are obvious; others contain fewer or smaller faces and may be less obvious. Zones can often be found by looking for parallel edges, although not all consecutive faces in a zone intersect, e.g. D and E in the zone BDEB'D'E'.

2. Different crystals of the same substance may be of a variety of shapes, but these are often sufficiently similar for one to be able to recognise *corresponding faces* as shown by a comparison of Fig. 1.1 and Fig. 1.2. There may be enormous differences between the relative sizes of corresponding faces, but it is found that the *angles* are always the same between pairs of corresponding faces on different crystals of the same substance. For

4   Crystallography for Earth Science Students

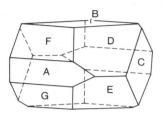

FIG. 1.2. A crystal with faces that are in a corresponding arrangement to those on Fig. 1.1 but of different relative sizes.

crystals of different substances the angles between faces may be so different that no such correspondence is evident.

These two features of crystal morphology, the zonal development of faces and the constancy of angles between corresponding faces, show that the faces of crystals do not occur at random or simply in directions where the crystal happens to stop. They constitute a very substantial restriction relative to all the possible facetted shapes that could be imagined, and this restriction must be related in some way to the internal structure of crystals. The zonal development of faces must be related to the principles of crystal structure in general, while the particular angles between corresponding faces on crystals of a particular mineral must be related to the specific structure of that mineral. Both features can be explained if we suppose that crystals are built up from stacks of sub-microscopic bricks. For the moment, we will not enquire into the physical reality of these sub-microscopic bricks, or how they relate to atomic or molecular structures; these questions are discussed in Chapter 2, after we have explored the usefulness of the model in explaining the basic facts of crystal morphology.

Figure 1.3(a) shows a stack of bricks with a variety of "faces" on it. Some of these are flat, being parallel to the faces of the bricks themselves; others are stepped, and yet others are

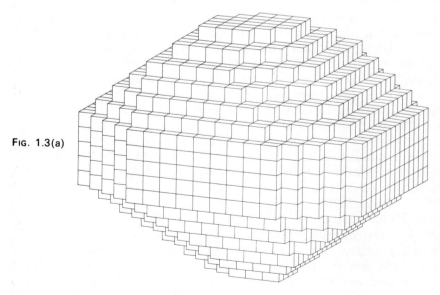

FIG. 1.3(a)

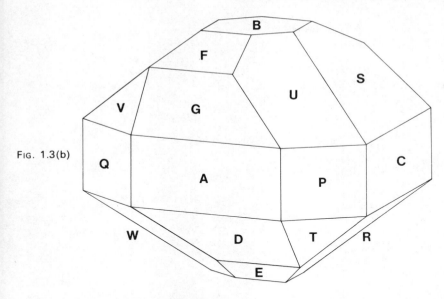

FIG. 1.3. (a) Stack of bricks showing flat, stepped and kinked faces. (b) The form that (a) would take up if the bricks were of the same shape but of infinitesimal size.

"kinked" with the corners of bricks sticking out of them. However, if the bricks were made smaller and smaller while retaining their shape, and the stack retained its total volume, the stepped and kinked faces would become smoother and smoother. When the bricks were too small to be seen the stack would have the appearance shown in Fig. 1.3(b), which displays the zonal development characteristic of a crystal. Zones that contain only faces which are flat or stepped in Fig. 1.3(a) are parallel to one or other of the edges of the bricks, because these form the edges of the steps. Zones that contain kinked faces have their zone axis parallel to various diagonals of the bricks.

It is natural to assume that crystals of any particular mineral will always be built of bricks of the same size and shape. The model then gives a ready interpretation of the existence of corresponding faces and the constancy of angles between them. Figure 1.4 shows a stack of bricks similar to that of Fig. 1.3 except that the stack has been extended in some directions and not in others. As a result some faces are larger and some are smaller, but correspondence between faces in the two figures is quite clear. The angles between the faces depend on the kind of steps involved; corresponding faces are those which are similarly stepped and the angles between them are necessarily equal.

Although the easiest kind of brick to imagine is a rectangular one, rather like an ordinary brick, this is not a necessary limitation. If the edges of the bricks are not at right angles to one another we can still build them into a stack having flat, stepped and kinked faces in zonal relationships to one another. Figure 1.5 shows such a stack built from bricks that have two rectangular faces and one oblique parallelogram face. The shape of the crystal is different of course; almost all the angles between the faces are different, and in particular the vertical flat faces of the stack are no longer at right angles. Stacks can equally

6   Crystallography for Earth Science Students

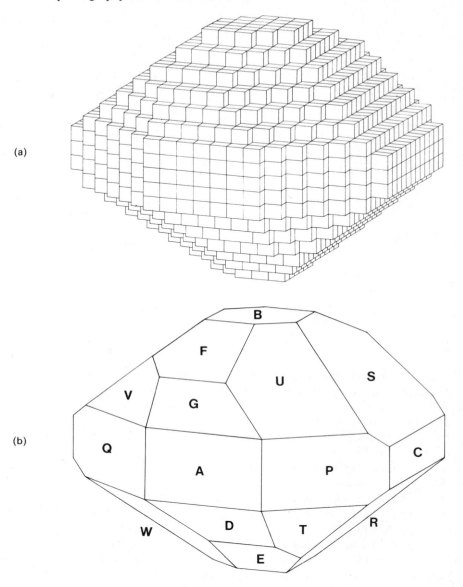

Fig. 1.4. Stack of bricks (a), and its smoothed form (b), as in Fig. 1.3, but extended outwards in some directions. As a result the faces are of different sizes but in corresponding directions.

well be built from bricks having all their edges oblique to one another, and in this case all the flat faces of the stack would be oblique to one another.

The concept of a crystal as a stack of submicroscopic bricks is of far more value than merely providing a qualitative explanation of zonal development and the constancy of angles between corresponding faces of a given mineral; it predicts quantitative relation-

External Form and Cell Structure of Crystals 7

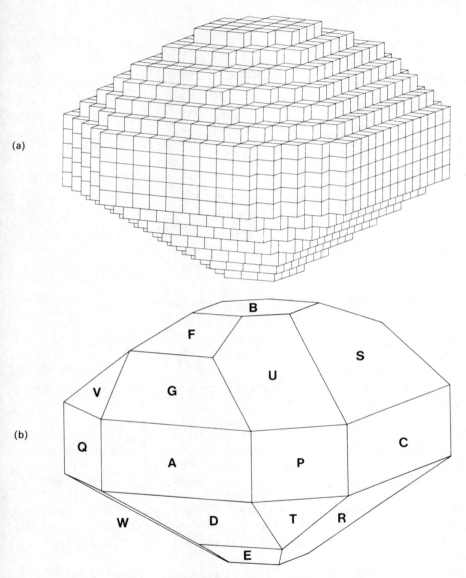

FIG. 1.5. Stack of bricks similar to Fig. 1.3(a) but built from non-rectangular bricks having vertical rectangular faces and horizontal parallelogram faces.

ships between various interfacial angles in a zone. From Fig. 1.6 it is evident that if we have faces that are stepped
  (i) 1 brick along × 2 bricks down (D),
 (ii) 1 brick along × 1 brick down (E),
(iii) 2 bricks along × 1 brick down (F),

8   Crystallography for Earth Science Students

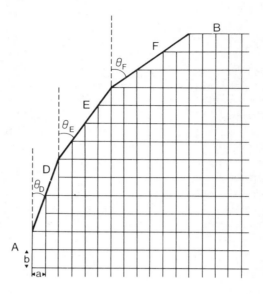

FIG. 1.6. Cross-section of a stack of bricks with three stepped faces of different kinds.

and the bricks are of length a and height b, then the angles θ_D, θ_E, θ_F that they make with the smooth vertical face A are given by

$$\tan \theta_D = a/2b, \ \tan \theta_E = a/b, \ \text{and} \ \tan \theta_F = 2a/b$$

respectively. Thus in general, if the steps are always constructed with small whole numbers of bricks along and down, then the ratio of the tangents of angles defined in this way will always be small whole numbers, and this is indeed found to be true for crystals.

This result means that we can find the shape of the bricks from which a crystal is built from the angles between its faces, provided that we can identify which faces are the "flat" ones (parallel to the faces of the bricks themselves) and the kind of steps on just two of the stepped faces.

We call the lengths of the edges of the bricks a, b and c, and we call the "flat" faces of the crystal that contain the bc, ca, ab faces of the bricks the A, B, C faces respectively. If we assume that some face in a zone between A and B (like G in Fig. 1.3) has steps one brick wide by one brick high, then we can calculate the ratio a/b from the angle between A and G as:

$$a/b = \tan A\hat{\ }G.$$

Similarly, we can calculate the ratio c/b if we assume that some face S in the zone between B and C is stepped one brick wide (in the c-direction) by one brick high, so that

$$c/b = \tan C\hat{\ }S.$$

The shape of the brick can then be expressed as $a:b:c = \tan A\hat{\ }G : 1 : \tan C\hat{\ }S*$. We are also then able to specify the relative height and widths of all the other stepped faces on the crystal by comparing the tangents of appropriate angles as in the discussion of Fig. 1.6.

The only problem therefore is to identify the "flat" faces A, B, C and two faces like E and G with unit steps. The appropriate way to set about this systematically cannot be specified fully until crystal symmetry has been discussed in Chapter 3, but a reasonable way to proceed in simple cases is as follows.

1. (i) Look for faces at right angles to one another. If there are three such faces that are mutually perpendicular then this suggests that the bricks are rectangular, and that these are the "flat" faces parallel to the faces of the bricks. The "flat" faces may then be assigned the symbols A, B, C in any order.

(ii) If three such faces cannot be found, but there is one face that is perpendicular to each of two others that make an angle somewhere near 90° to one another (e.g. in the range from 60° to 120°), then these can probably be taken as the "flat" faces. It is then conventional to call B the one that is perpendicular to the other two (A and C). The bricks are then of the shape shown in Fig. 1.7; the edge b is perpendicular to a and c, which make an angle of β with one another.

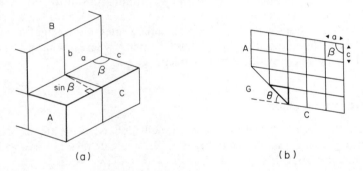

FIG. 1.7. The shapes of steps on stacks of non-rectangular bricks that have b perpendicular to a and c. (a) Perspective view of a step on a face in the zone between the B and C faces. (b) Cross-sectional view of steps on a face in the zone between the A and C faces.

(iii) If there are no perpendicular faces the bricks may have all their edges obliquely inclined to one another, but consideration of this case will be postponed till later.

2. If the bricks do not have very unequal edge lengths, then the faces with unit steps will make angles with the "flat" faces that are not very far removed from 45°. Find two such faces, which may be in any two of the three zones between A–B, B–C or C–A. On the assumption that these are indeed faces with unit steps one can calculate $a:b:c$ directly if the bricks have been found to be rectangular.

---

* The ratios $a:b:c$ are called the *axial ratios* of the crystal, and it is conventional to represent them in this form in which b is set arbitrarily equal to unity so that $a:b:c = \frac{a}{b} : 1 : \frac{c}{b}$.

If the bricks are not rectangular, but of the shape discussed in (ii) above, it is necessary to allow for the fact that the step widths in the zone between A and B are $a \sin \beta$ instead of a (see Fig. 1.7(a)) and in the zone between B and C they are correspondingly $c \sin \beta$ instead of c; also in the zone between A and C the steps themselves are not rectangular, although the widths are a and c (Fig. 1.7 (b)). If a face of this kind is used to find $a:c$ then a general triangle instead of a right-angled triangle has to be solved, the appropriate triangle being shown by the bold lines in Fig. 1.7(b). The value of β is obtained directly from the angle between the faces A and C.

Once this process is complete the kinds of steps can be found on all the stepped faces in all three zones between the flat faces, i.e. A–B, B–C, C–A. In principle it is also possible to find the kind of kinks present on the kinked faces, but this will not be attempted until a more convenient notation has been introduced in Chapter 4.

The process described above may well seem to be a rather unsatisfactory basis on which to build the science of crystallography. We are on fairly safe ground in identifying the faces A, B and C provided that they are mutually perpendicular; but it might be argued that it is quite arbitrary to assume that the oblique angle β of a non-rectangular brick will necessarily be near to 90°, and if it is not there would seem to be no particular way to justify our choice of the faces A and C from among the zone of faces that are all perpendicular to face B in Fig. 1.5. Equally it could be argued that we cannot know in advance that the edges a, b, c will not be very unequal, and therefore that we have no logical basis for assuming that the faces with unit steps will be those which make angles in the neighbourhood of 45° with the "flat" faces. These questions are left to Chapter 2, where it is shown that there are reasons why they do not constitute real problems at all.

Conventions regarding angles

The interfacial angles of crystals are conventionally defined as the angles between their outward-pointing normals, as shown in Fig. 1.8. The angle defined in this way is equal to the *external* dihedral angle between the faces.

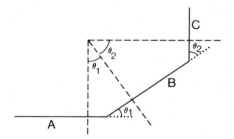

FIG. 1.8. Cross-section of a crystal perpendicular to a zone of faces (A, B, C . . .). Angles between faces (θ_1, θ_2 . . .) are defined as the angles between their normals (– – –). They are equal to the external dihedral angles as shown.

Interfacial angles on large crystals, and on crystal models, are measured with a contact goniometer. This is an ordinary 180° protractor with an arm centrally pivoted at the origin of the protractor. When the arm is in contact with one face and the base of the protractor

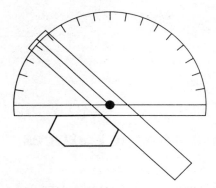

Fig. 1.9. Contact goniometer in use to measure an interfacial angle on a crystal whose cross-section is shown by the bold lines. The goniometer would be graduated in degree intervals but these are omitted.

with another, the index end of the arm indicates the value of the external dihedral angle, and therefore of the interfacial angle (Fig. 1.9), provided that the plane of the protractor is held perpendicular to both the faces. The accuracy obtainable is of the order of 1° at best.

For small crystals with faces less than a few mm across it is not practicable to use a contact goniometer. Interfacial angles are then usually measured with an optical goniometer; a beam of light is reflected in a chosen direction from one face and the crystal is then turned until the beam is reflected in exactly the same direction by another face. The angle through which the crystal has to be turned is the angle between the two faces. If the faces are of good quality (bright and smooth) the accuracy can be as good as 1' of arc.

Because the faces in a zone are all parallel to the zone axis, their normals all lie in a plane perpendicular to the zone axis. The sum of all the angles between consecutive faces in a zone is therefore always 360°. It is very common (though not an invariable rule) for crystals to develop pairs of opposite faces which are parallel to one another. Such faces are present on all the illustrations of crystals in this chapter, and it follows from the definition of interfacial angles in terms of the angle between outward pointing normals that the angle between any pair of opposite faces is 180°.

Problems

1. On the crystal shown in Fig. 1.3(b) some interfacial angles are $A\hat{\ }P = 32°$, $P\hat{\ }C = 58°$, $A\hat{\ }G = 37\frac{1}{2}°$, $G\hat{\ }B = 52\frac{1}{2}°$, $B\hat{\ }C = 90°$. Assume that the faces A, B, C are perpendicular to the edges of the unit cell that are of length a, b, c, respectively. If the faces are stepped as shown in Fig. 1.3(a) calculate the axial ratios $a:b:c$.
2. From the axial ratios found in question 1, calculate the interfacial angles $F\hat{\ }B$ and $C\hat{\ }S$ on the basis of the stepping shown in Fig. 1.3(a).
3. On the crystal shown in Fig. 1.5(b) some interfacial angles are $A\hat{\ }P = 26°$, $P\hat{\ }C = 45°$, $A\hat{\ }G = 37°$, $G\hat{\ }B = 53°$, $B\hat{\ }C = 90°$. Assume that the faces are stepped as shown in Fig. 1.5(a). Calculate the interaxial angles α, β, γ and the axial ratios $a:b:c$.
4. From the interaxial angles and axial ratios found in question 3 calculate the interfacial angles $F\hat{\ }B$ and $C\hat{\ }S$ on the basis of the stepping shown in Fig. 1.5(a).

CHAPTER 2

Repeating Patterns and Lattices

THE concept developed in Chapter 1, that crystals have the shape to be expected if they are built of stacks of submicroscopic bricks, requires amplifying in terms of our knowledge that crystals, like all other kinds of matter, must be built from some arrangement of atoms. The two concepts are in fact compatible providing that the atoms in a crystal are arranged in a repetitive pattern. To elucidate this we must consider the characteristics of repetitive patterns, and for simplicity we start by considering two-dimensional examples such as are frequently seen on wallpapers and linoleum.

An example of a two-dimensional repetitive pattern is shown in Fig. 2.1(a). If an arbitrary point is chosen anywhere in this pattern (as at ×), then it is possible to find a whole array of such points (Fig. 2.1(b)) all of which are exactly similarly situated with respect to the pattern. This array of points is shown without the pattern in Fig. 2.1(c). It

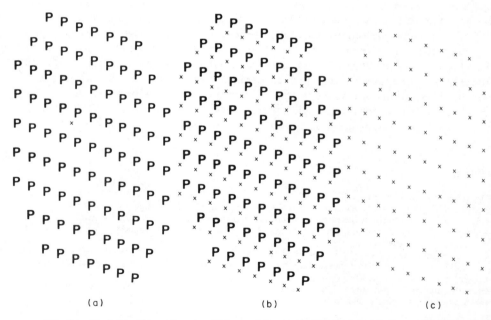

FIG. 2.1. (a) A repeating pattern in which an arbitrary point has been inserted at ×. (b) Points that are exactly similarly situated inserted throughout the pattern. (c) The lattice of points describes the repetitive characteristics of the pattern in (a).

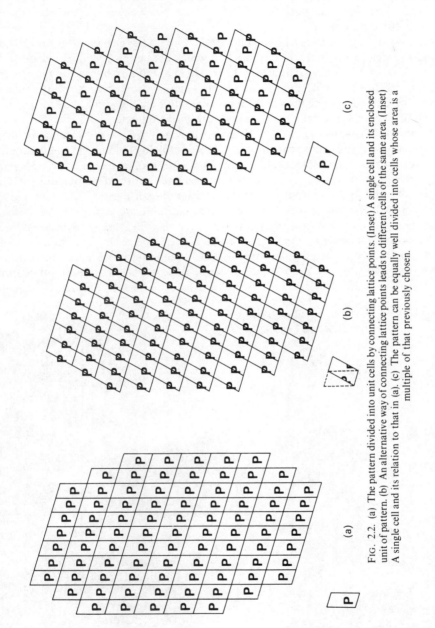

FIG. 2.2. (a) The pattern divided into unit cells by connecting lattice points. (Inset) A single cell and its enclosed unit of pattern. (b) An alternative way of connecting lattice points leads to different cells of the same area. (Inset) A single cell and its relation to that in (a). (c) The pattern can be equally well divided into cells whose area is a multiple of that previously chosen.

gives a complete description of the repetitive character of the pattern, abstracted from the substance of it. An exactly geometrically similar lattice would have been obtained wherever we had placed the initial point in Fig. 2.1(a). Such an array of points is the *lattice* of the repetitive pattern.

If we now join up the lattice points on Fig. 2.1(b) by straight lines as in Fig. 2.2(a) then the whole pattern is divided into cells, each of which contains a unit of the pattern. These cells are called *unit cells*. The pattern can be regarded as built up by stacking together such unit cells, each containing its unit of pattern. The edges of the pattern may lie parallel to the edges of the unit cells, or they may be at an angle to them, in which case the edge of the pattern corresponds to a set of steps in the array of unit cells.

Thus the submicroscopic bricks, in terms of which we have interpreted the zonal development of crystals, may in turn be interpreted as being the unit cells of the pattern of atoms from which crystals are really built. This pattern of atoms is a *three-dimensional* repeating pattern; it therefore has a three-dimensional lattice of points all of which are similarly situated in the pattern, and if the lattice points are joined up by straight lines these lines will outline a brick-shaped three-dimensional unit cell. Whereas two-dimensional repeating patterns are familiar in everyday life, three-dimensional ones are much less common, but an example is shown in Fig. 2.3. These water pots on the banks of the Nile are stacked up in a regular three-dimensional pattern each unit of which consists of a pair of water pots arranged "head-to-tail". If the unit cells containing these units of

Fig. 2.3. Water-pots stacked in a three-dimensional repeating pattern. The stacks have "crystal faces".

pattern were to be drawn in, the stacks would be seen to have flat, stepped and kinked faces in terms of the brick-shaped unit cells, just like the stacks of bricks in Figs. 1.3–1.5. However, even without drawing in the unit cells, the zonal development of faces is evident on these stacks of water pots, and the ways in which the pots are exposed on different faces correspond directly to the ways in which atoms are exposed on the faces of crystals.

Identification of the submicroscopic bricks with the unit cells of repeating patterns resolves the difficulties that faced us in Chapter 1 over the seemingly arbitrary choices that we had to make in finding the ratios $a:b:c$ of the edge-lengths of the bricks. The most crucial of these choices was that of deciding which faces of the crystal were to be taken as being parallel to the faces of the bricks. If there were faces at right angles to one another (as in Fig. 1.3) it seemed appropriate to choose these, but when this criterion failed (as in Fig. 1.5) we chose faces that corresponded to bricks that were as nearly rectangular as possible. If the bricks were real material entities then this would have amounted to a guess, which might have been wrong. However, now that we have identified the "bricks" with the unit cells of a repeating pattern the whole problem disappears. There are many alternative ways of joining up the points of a lattice to outline unit cells, as shown in Fig. 2.2, and none of these is wrong. Any such choice of unit cell can be regarded as containing a unit of pattern, even though this unit may be built up of complementary parts of what we would at first sight take as adjacent units of the pattern; the whole pattern can be built up equally well by stacking together the units of pattern defined by any of the possible choices of unit cell. Thus there is no right or wrong choice of crystal faces in defining the faces of the submicroscopic bricks; there is just a set of alternative choices in defining the unit cell of the crystal, some of which will be more convenient than others. In particular it will always be most convenient to choose a rectangular cell if we can, and if we cannot it will usually be most convenient to choose one that is as nearly rectangular as possible.

It may be seen from Fig. 2.2 that the edge-lengths of different choices of unit cell are always related to one another; an edge of one choice will be a diagonal of another, and a diagonal of a combination of two or more cells of yet another.

In addition to the choice of which crystal faces are to be taken as parallel to the faces of the unit cell, there was another seemingly equally arbitrary choice that had to be made in Chapter 1. This was the choice of which faces we should assume to have unit steps of one brick out by one brick down. There is a sense in which this is a more serious problem than the previous one, inasmuch as once the unit cell edges of a repeating pattern have been defined the true unit cell has a definite size. Thus if we assume unit steps to exist on a crystal face where in fact there are multiple steps of some kind, then we shall obtain the wrong values of $a:b:c$. However, the effect will merely be that one or two of the edges of the cell will be multiplied by a small whole number, which will very rarely be more than 2 or 3; and although such a multiple of the true unit cell will contain more than one unit of pattern, it will nevertheless have the characteristic, in common with a true unit cell, that the whole pattern can be generated by stacking together such "units", and it will lead to a self-consistent description of the kinds of steps present on all the faces of the crystal. Thus if in Fig. 2.2(b) the top left and top edges of the pattern are taken to be parallel to the edges of the unit cells as shown, but the top right edge were assumed to be stepped one cell down by one cell out, this would correspond to the choice of a "unit" cell that was doubled in one direction, as shown in Fig. 2.2(c). In terms of this unit cell the vertical edges of the pattern would be deduced from their angular relationships to be stepped two cells down by one cell out. This is obviously not a very good description, because it makes the steps unduly open and ignores the presence of Ps that partially fill up these open steps. Nevertheless, if the scale of the pattern were submicroscopic we would not be aware of this, and the choice of cell shown in Fig. 2.2(c) would give a perfectly self-consistent account of all the angular relationships between the edges of the pattern. It is not possible on the basis of morphological crystallography to obtain certainty as to whether we have obtained a true unit cell or a multiple of it. Such certainty can only be obtained by X-ray methods, as we shall see in Part II. From morphological considerations we can only work on a balance of probabilities, and in absence of contrary evidence take those faces to have unit steps which lead to $a:b:c$ ratios nearest to unity. Contrary evidence would be if some other face were consistently more prominent on the crystals, or especially if the crystals had a prominent cleavage* parallel to some other face: this would be good evidence that that face had unit steps.

Problems

1. In a real crystal we would not be able to see the unit cells as in Fig. 1.3(a) and this stepping of the faces would be a pure assumption. Make the different assumption that face F is stepped $a \times b$ instead of $2a \times b$, but with the same assumption as before about the stepping of P. Recalculate the axial ratios, using the angle $F\widehat{\ }B$ found in question 1 of Chapter 1 (p. 11).
2. How is the cell that results from the assumptions in question 1 related to that shown in Fig. 1.3(a)? What kind of steps would then be present on G and S?
3. Calculate the interfacial angle $A\widehat{\ }Q$ in Fig. 1.5(b).
4. For the crystal shown in Fig. 1.5(b) make the assumption (contrary to the disposition of unit cells shown in Fig. 1.5(a)) that the face Q is parallel to the bc face of the unit cell and that the face A is stepped $a \times c$, but retain the previous assumptions about faces B, C and S. Some interfacial angles are $Q\widehat{\ }A = 36°$, $A\widehat{\ }P = 26°$, $P\widehat{\ }C = 45°$, $A\widehat{\ }G = 37°$, $G\widehat{\ }B = 53°$, $C\widehat{\ }S = 50\frac{1}{2}°$, $S\widehat{\ }B = 39\frac{1}{2}°$. Find the new values of α, β, γ and $a:b:c$ on these assumptions.
5. How is the cell that results from the assumptions in question 4 related to that shown in Fig. 1.5(a)? What kind of steps would then be present on face P?

---

* Many, but not all, crystals break cleanly along planes parallel to one or more crystal faces. This is a phenomenon called cleavage and arises from the existence of planes, through the repeating pattern of atoms, across which the interatomic bonding is minimal.

CHAPTER 3
Crystal Symmetry

WE SAW in Chapter 1 that in order to deduce the form of the unit cell of a crystal from its interfacial angles we had to choose which faces of the crystal we would assume to be parallel to the faces of the unit cell. In order to make an intelligent choice in this respect we need to know what shapes of unit cell are possible, and how to recognise from the appearance of a crystal what type of unit cell to expect it to have. For this purpose unit cells are classified according to their symmetry (or more precisely the symmetry of the lattice which arises from stacking them together), because this symmetry is reflected in the external symmetry of the crystal. Conversely, therefore, by contemplating the symmetry of a crystal we can decide what type of unit cell it must have, and so know how best to set about finding it.

Figure 3.1 shows a number of shapes which have recognisable kinds of symmetry. A *symmetry operation* moves or transforms an object in such a way that after transformation it coincides with itself. Thus the object in Fig. 3.1(c) coincides with itself after the operation of rotation through 120° or 240° about its centre; it has 3-fold symmetry because it coincides with itself three times during a complete rotation and the point at its centre is called a 3-fold rotation point. Similarly Fig. 3.1(a) coincides with itself after reflection in a line, which is called a mirror line; it is said to have mirror symmetry. The mirror line, and the points of rotation symmetry of various kinds in Fig. 3.1, are *elements of symmetry* applicable to two-dimensional shapes, and they are conveniently denoted by *m* (for mirror) and 2, 3, 4 and 6 (the order of their symmetry).

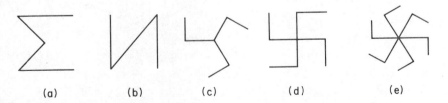

FIG. 3.1. Shapes that possess two-dimensional symmetry elements: (a) a mirror line, *m*; (b)-(e) rotation points of order 2, 3, 4 and 6.

Three-dimensional objects have corresponding elements of symmetry, though in three dimensions reflection takes place in a *mirror plane* (again denoted by *m*) and rotation about *axes of rotation* (again denoted by 2, 3, 4, 6). In three dimensions there are, however, some

18   Crystallography for Earth Science Students

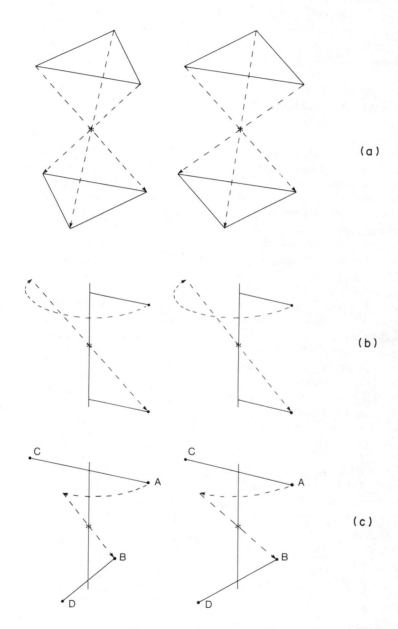

Fig. 3.2. Stereoscopic drawings showing the symmetry operations involving inversion: (a) a centre of symmetry ($\bar{1}$) acting on a triangle; (b) a $\bar{2}$-axis of rotation–inversion, equivalent to a perpendicular mirror plane; (c) $\bar{4}$-axis of rotation–inversion moves A to B, B to C, C to D and D to A. (Instructions for viewing stereoscopic drawings are given on p. 27.)

additional kinds of symmetry. The simplest is the *centre of symmetry* illustrated in Fig. 3.2(a). If there is a centre of symmetry at the origin, the corresponding symmetry operation transforms an object by taking each point of it (at x, y, z) to a corresponding point (at $-x, -y, -z$) an equal distance at the opposite side of the centre of symmetry. Such a transformation is called an *inversion* in the centre. The other additional kinds of symmetry in three dimensions are obtained by combining inversion with each kind of rotational symmetry to give *axes of rotation–inversion*. The symmetry operation involved in 4-fold rotation–inversion symmetry is rotation through 90° about the axis *plus* inversion through the origin (Fig. 3.2(c)), and similarly for rotation–inversion symmetry of other orders. Such symmetry elements are denoted $\bar{2}, \bar{3}, \bar{4}$ and $\bar{6}$, but it may be seen from Fig. 3.2(b) that the symmetry operation of $\bar{2}$ is equivalent to reflection in a mirror plane through the centre and perpendicular to the axis, and the notation m is usually preferred to $\bar{2}$. By extension of the nomenclature the centre of symmetry is denoted $\bar{1}$, and a total lack of symmetry can be denoted 1 (corresponding to inevitable identity after rotation through a full 360° turn).

In diagrams it is convenient to represent the rotational symmetry elements by a figure of the appropriate number of sides, filled for ordinary rotation and empty for rotation–inversion, i.e. ♦, ▲, ♦, ●, △, ◇, ○. The centre of symmetry is denoted by a very small open circle, ○, and the edge of a mirror plane by a bold line.

The symmetry elements 1, 2, 3, 4, 6, $\bar{1}, \bar{2},$ (= m), $\bar{3}, \bar{4}$ and $\bar{6}$ are the only ones which are found in the morphology of crystals and they are known as the crystallographic symmetry elements. Of course symmetry axes of any integral order (e.g. 5, or $\bar{7}$, etc.) can be devised and objects can be constructed which possess such non-crystallographic symmetry. However it may be proved mathematically that repetitive patterns cannot possess elements of non-crystallographic symmetry, a fact which is connected with the impossibility of fitting together regular pentagons or heptagons (or higher-order polygons) without leaving spaces between.

Symmetry of unit cells in two dimensions

The unit cell of a plane repeating pattern is necessarily a quadrilateral with equal and parallel opposite sides. Its minimum symmetry is therefore that of a parallelogram (Fig. 3.3(a)) which has 2-fold rotation symmetry. If mirror symmetry is added, with the mirror line perpendicular to one of the edges, then it becomes a rectangle (Fig. 3.3(b)). A second mirror line is inevitably introduced perpendicular to the other edge. Higher symmetry can be achieved by converting the 2-fold rotation point to a 4-fold rotation point, when the unit cell becomes a square (Fig. 3.3(c)) and diagonal mirror lines are inevitably introduced. The unit cell itself can have no higher symmetry than this, but an

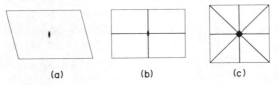

FIG. 3.3. Symmetry of two-dimensional unit cells with increasing symmetry: (a) parallelogram, (b) rectangle, (c) square.

20   Crystallography for Earth Science Students

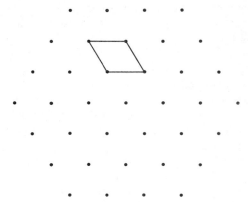

Fig. 3.4. A lattice with hexagonal symmetry showing the shape of its unit cell.

array of unit cells can have 6-fold symmetry. The unit cell then has its shorter diagonal equal to both its sides (Fig. 3.4), and it is therefore a rhombus with 60° and 120° angles.

There are therefore four kinds of symmetry possible in two-dimensional repeating patterns, and such patterns can be classified according to their symmetry as oblique, rectangular, square or hexagonal.

Symmetry of unit cells in three dimensions

The development here is a little more complicated but quite analogous. Any three-dimensional unit cell is a parallelepiped (pronounced parallel'epīped with a short i), a solid figure having six parallelogram faces, parallel to one another in pairs. In its most general form it therefore has a centre of symmetry ($\bar{1}$) at its centre, and no other symmetry (Fig. 3.5(a)). It has three unequal edges of lengths a, b, c which make unequal angles α, β, γ with each other, and none of these angles is a right angle. It is conventional to take α to be the angle between b and c, β between c and a, and γ between a and b. Crystals in which it is not possible to find a unit cell with any more symmetry than this are said to belong to the *triclinic system*, since all *three* cell edges are *inclined* to one another. It is also sometimes called the *anorthic system* because it contains no right angles.

The parallelepiped may be made more symmetrical by putting a mirror plane perpendicular to one set of four parallel edges, conventionally those of length b. The other edges (a and c) then have to be parallel to the mirror plane, so the angles α and γ become 90° (Fig. 3.5(b)). In addition to the mirror plane there is now a 2-fold rotation axis parallel to the edge b. Crystals in which it is possible to find a unit cell whose shape has this symmetry (but not possible to find a unit cell of higher symmetry) are said to belong to the *monoclinic system*, because the cell has *one inclined* angle.

The process may be continued by inserting a second mirror plane perpendicular to a second set of edges. This makes all the angles of the unit cell equal to 90°, and inevitably introduces a third mirror plane perpendicular to the third set of edges, and 2-fold rotation

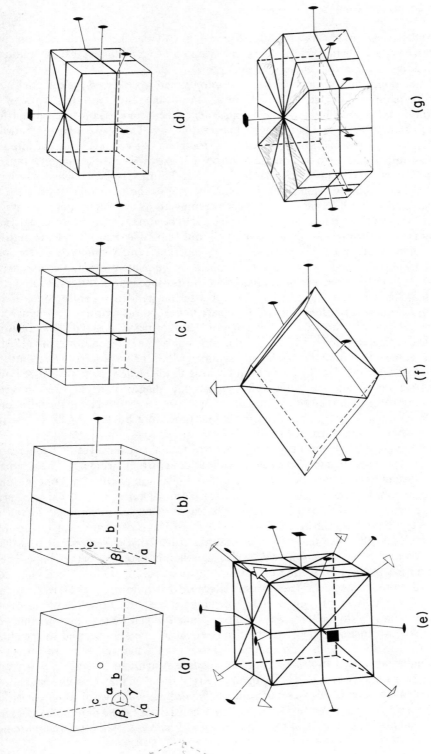

Fig. 3.5. Symmetry of three-dimensional unit cells: (a) triclinic, (b) monoclinic, (c) orthorhombic, (d) tetragonal, (e) cubic, (f) rhombohedral, and (g) a composite group of two cells and two half-cells having hexagonal symmetry.

axes parallel to all three sets of edges (Fig. 3.5(c)). All the faces of the cell are now rectangles. Crystals in which a unit cell of this symmetry (but no higher) can be found are said to belong to the *orthorhombic system*.

Up to this point the symmetry introduced into the cell has affected the cell angles but not the inequality of the cell edges. If now one of the 2-fold rotation axes of an orthorhombic cell (conventionally the one parallel to the c edge) is converted to a 4-fold rotation axis, then the ab face becomes a square with $a = b$. This also leads automatically to the insertion of two additional diagonal mirror planes at 45° to those that intersect along the 4-fold axis, and to diagonal 2-fold rotation axes where these diagonal mirrors intersect the mirror plane perpendicular to them (Fig. 3.5(d)). Crystals in which a unit cell of this symmetry (but no higher) can be found belong to the *tetragonal system*.

If we take an orthorhombic cell and, while keeping all its existing symmetry, insert a 3-fold axis along a body diagonal, then all three edges become symmetrically equal and a cube results. This then has 3-fold axes along all four body diagonals, which are actually axes of rotation–inversion ($\bar{3}$) because of the inherent centrosymmetry of the parallelepiped. This happens here because a $\bar{3}$-axis is equivalent to a 3-axis plus an independent $\bar{1}$-centre. The corresponding effect did not occur in the tetragonal system because a $\bar{4}$-axis is not equivalent to a 4-axis plus an independent $\bar{1}$-centre. Since all the faces are now squares all three 2-fold rotation axes of the orthorhombic cell are converted to 4-fold rotation axes; and six additional diagonal mirror planes and six 2-fold diagonal rotation axes are introduced (Fig. 3.5(e)) in the same way as the two additional ones of each in the tetragonal unit cell. This is the highest symmetry that can be possessed by a unit cell, and crystals in which such cells can be found belong to the *cubic system*. It is to be noted here that we developed the cubic cell by inserting diagonal 3-fold axes into the orthorhombic cell, rather than converting additional 2-fold to 4-fold axes in the tetragonal cell. The latter might appear to be the more obvious procedure, but the former is preferred for reasons that will appear in Chapter 7.

In developing these five systems starting from the triclinic we have gradually increased the orthogonality, and then maintained this while adding further symmetry. There remain two other systems which must be developed differently and in parallel with one another.

A 3-fold axis may be put along one body diagonal only of a triclinic cell. This makes all three edges equal and all three angles equal, but leaves them non-orthogonal. Again the 3-fold axis combines with the inversion centre to give a $\bar{3}$-axis of rotation–inversion. The shape of the unit cell is like that of a cube that has either been pulled out, or squashed, along one diagonal. The well-known calcite cleavage rhombohedron is of this shape. There are three mirror planes intersecting at 120° in the $\bar{3}$-axis, and three 2-fold rotation axes joining mid-points of opposite "equatorial" edges of the rhomb (Fig. 3.5(f)).

Six-fold symmetry cannot be inserted into a parallelepiped, but a stack of unit cells can have 6-fold symmetry (in the same way as in two dimensions) if we take a monoclinic cell and make the two inclined edges symmetrically equal to one another and to the short diagonal of the face which they bound. The perpendicular edge in this case is conventionally taken as c (not b as in the monoclinic system) so that $\alpha = \beta = 90°$ and $\gamma = 120°$. The symmetry of a lattice having unit cells of this type may be appreciated if we recollect that a lattice consists of a set of points only; the lines which we draw in to indicate the unit cell are figments of our imagination. The whole lattice can be built up by stacking together hexagonal prism units of the type shown in Fig. 3.5(g). Although these are not unit cells (they include six symmetrically related half-unit cells) they possess all the symmetry of the lattice. Their symmetry elements are a 6-fold rotation axis with a mirror plane perpendicular to it, six 2-fold rotation axes through the centres of the prism faces and the mid-points of the vertical edges, and six mirror planes perpendicular to each of these 2-fold axes.

The definition of crystal systems in terms of these rhombohedral and hexagonal unit cells is fraught with difficulties and has been the subject of controversy, because lattices based on them are closely related. In a lattice based on a stack of rhombohedral cells one can find a larger cell having the *shape* of the hexagonal cell, but with two lattice points within it which destroy the hexagonal symmetry. Equally in a stack of hexagonal cells one can find a larger rhombohedral cell having two lattice points within it. Because of this close relationship some authors define a single sixth system embracing both types of lattice. Others define rhombohedral and hexagonal systems based strictly on the shape of their *smallest* unit cell, but this has the disadvantage that it is not always possible to tell from the external symmetry of the crystal which of the two kinds of lattice it possesses. Accordingly in this book we adopt the approach that is based most closely on the directly observable symmetry of the crystal. If a crystal possesses a 6-fold rotation (or rotation–inversion) axis it belongs to the *hexagonal system*, and if it possesses a 3-fold rotation (or rotation–inversion) axis, but not a 6-fold one, it belongs to the *trigonal system*. Further discussion of the relationships between these two systems will be found in Chapters 6 and 11.

The minimum symmetry characterising a crystal system

A summary of the characteristics of the *seven crystal systems* is given in Table 3.1. The definitions of the first six systems can be made to depend on the symmetry of the shape of the unit cell and this is also the symmetry of the lattice regarded as extending indefinitely in all directions, whereas this latter criterion alone is true for the hexagonal system. However, it is possible for a repeating pattern to have a lower symmetry than that of its lattice. This has already been demonstrated by the two-dimensional pattern in Fig. 2.1; this has no symmetry because of the unsymmetrical character of the motif (the letter P), but the shape of the unit cell, and the lattice, have 2-fold rotational symmetry. Figure 3.6 shows a pattern that has exactly the same lattice, but in which the motif (the letter N) also possesses 2-fold rotational symmetry. The infinitely extended pattern then has the same symmetry as its lattice, and it is possible to choose the lattice points in an appropriate relationship to the motif so that the unit cell contents have the same symmetry as the unit cell shape. In exactly the same way, in three dimensions the contents of a triclinic cell may either have no symmetry at all, or they may have the same symmetry as the shape of the cell, namely a centre of inversion ($\bar{1}$).

In the triclinic system this raises no problems, but in systems of higher symmetry the question arises as to what is the minimum symmetry that a repeating pattern may have if its lattice is to have the symmetry appropriate to the system in question. This problem may be approached most simply in the monoclinic system which we derived from the triclinic on p. 20 by setting a mirror plane perpendicular to the b-edge of the unit cell, thereby causing the a and c edges to become perpendicular to b, and we noted that the shape of the

TABLE 3.1 *The seven crystal systems*

Crystal system	Cell shape	Symmetry of shape of cell	Cell axes	Characteristic symmetry*	No. of classes in the system
Triclinic	General parallelepiped	$\bar{1}$	$a \neq b \neq c$ $\alpha \neq \beta \neq \gamma \neq 90°$	monad	2
Monoclinic	Right prism on parallelogram as base	$2/m$	$a \neq b \neq c$ $\alpha = \gamma = 90° \neq \beta$	1 diad	3
Orthorhombic	Rectangular parallelepiped	$2/m\ 2/m\ 2/m$	$a \neq b \neq c$ $\alpha = \beta = \gamma = 90°$	3 diads	3
Tetragonal	Square prism	$4/m\ 2/m\ 2/m$	$a = b \neq c$ $\alpha = \beta = \gamma = 90°$	1 tetrad	7
Cubic	Cube	$4/m\ \bar{3}\ 2/m$	$a = b = c$ $\alpha = \beta = \gamma = 90°$	4 triads	5
Trigonal	Cube deformed along one diagonal	$\bar{3}\ 2/m$	$a = b = c$ $\alpha = \beta = \gamma \neq 90°$	1 triad	5
Hexagonal	Prism on 60°–120° base	This cell does not itself have hexagonal symmetry, but the complete lattice does	$a = b \neq c$ $\alpha = \beta = 90°\ \gamma = 120°$	1 hexad	7

* The characteristic symmetry is the minimum required to assign a crystal to a particular system. More symmetry than this is possible up to the "symmetry of shape of cell"; hence the various different crystal classes in each system.

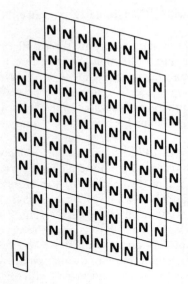

FIG. 3.6. Repeating pattern with the same lattice as Fig. 2.1 but with a motif having 2-fold rotational symmetry. (Inset) A choice of cell such that its contents have the same symmetry as its shape.

cell that we obtained had a 2-fold rotation axis parallel to b. Now the introduction of the mirror plane necessarily required the contents of the cell to have mirror symmetry, but it did not require opposite sides of the cell to be symmetrically related—it could contain a solid P-shaped object as in Fig. 3.7(a). Thus a mirror plane alone is sufficient to make the cell monoclinic. On the other hand, we could alternatively have derived the monoclinic cell from the triclinic by putting a 2-fold rotation axis parallel to the b-edge; this would have been equally effective in making the other edges perpendicular to b, but would not have required the opposite ends of the cell to be symmetrically related—it could contain a solid N-shaped object black at one end and white at the other as in Fig. 3.7(b). Thus the minimum symmetry of the monoclinic system is *either* a mirror plane *or* a 2-fold rotation axis. Since a mirror plane can equally well be regarded as a 2-fold axis of rotation–inversion, it is convenient to define the term *diad* to mean either kind of 2-fold axis, and so to define the minimum symmetry of the monoclinic system as one diad. The

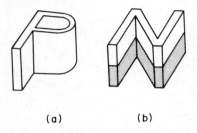

FIG. 3.7. Solid shapes having (a) a horizontal mirror plane and (b) a vertical 2-fold axis but no mirror plane.

26   Crystallography for Earth Science Students

terms *triad*, *tetrad* and *hexad* will be used similarly to mean either kind of 3-fold, 4-fold and 6-fold axis respectively.

From the foregoing it can be seen that there are three different types of symmetry which a crystal may have and yet belong to the monoclinic system; a mirror alone, a 2-fold axis alone, or both together. Crystals having the same type of symmetry are said to belong to the same *crystal class*, and there are therefore three crystal classes in the monoclinic system. In each crystal system there are several crystal classes, and they range in symmetry from the minimum required to define the shape of the unit cell of the system to the full symmetry of the lattice.

The detailed development of all the crystal classes (of which there are thirty-two) will be dealt with in Chapter 7. The purpose of the present chapter is to provide a logical basis for choosing an appropriate unit cell of any given crystal, and for this purpose we need only to be able to recognise unambiguously to what system it belongs. To do this we need to know the minimum symmetry that is characteristic of each system (i.e. the symmetry of the least symmetric crystal class in the system) which is listed in the last column of Table 3.1. Then, with an unknown crystal, if one deduces its symmetry by contemplating its external form, this must be in the range from the minimum to the maximum symmetry of a system. In all cases it will suffice to determine the system, and, armed with this knowledge, we can resolve many of the uncertainties and arbitrary choices which seemed to face us in Chapter 1 in choosing the crystal faces that we would assume to be parallel to the faces of the unit cell. Thus the crystal illustrated in Fig. 1.3 has three diads perpendicular to the faces A, B, and C; so it is orthorhombic and must have a rectangular unit cell with its edges parallel to these diads, and therefore with its faces parallel to A, B and C. This was the assumption that we made for the sake of convenience in Chapter 1, but we can now see that it was necessary. Similarly the crystal illustrated in Fig. 1.5 has a diad perpendicular to face B, but no others. The crystal is therefore monoclinic, and its unit cell must have its *b*-edge parallel to the diad and its *a*- and *c*-edges lying in the face B; their specific directions remain a matter for relatively arbitrary choice.

It may seem from this reconsideration of Figs. 1.3 and 1.5 that rather little has been gained from a somewhat lengthy discussion of crystal symmetry, since it leads only to the same conclusions that we reached on more intuitive grounds in Chapter 1. However, many crystals are not susceptible to the intuitive type of consideration. The crystal depicted in Fig. 3.8(a) has no faces perpendicular to one another, but it has three diads at right angles.

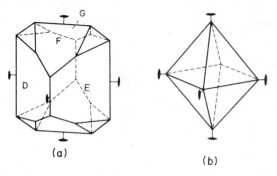

FIG. 3.8. Two orthorhombic crystals that have no face parallel to a face of the unit cell.

It is therefore orthorhombic and must have a rectangular unit cell with its edges parallel to these diads, although there is no crystal face parallel to any face of such a unit cell. However, from the symmetry we can deduce the orientation of the cell; we can then recognise that the edge between faces D and E is parallel to an edge of the cell, and similarly for the edge between F and G; and hence we know that these faces are all simply stepped and from the angles between them we can determine the axial ratios. Without a consideration of the symmetry the only clue we could have got from a crystal like this would have been from the fact that the edges between D and E and between F and G are at right angles. But it is difficult to measure angles between crystal edges, and with a crystal like that shown in Fig. 3.8(b) even this clue would be denied us. This crystal also has three diads and is therefore orthorhombic, but it has no face and no edge parallel to any face or edge of the unit cell. Thus all the faces are kinked faces, and we have so far fought shy of the problems involved in using kinked faces to calculate axial ratios. Before tackling such problems we need to develop a better way of describing the relationship of various kinds of faces to the unit cell, and this is done in the next chapter.

It is important to note here that when we speak of the symmetry of the shape of the unit cell, we are concerned only with symmetry that is imposed jointly by the fact that the unit cell must be a parallelepiped and by the symmetry of the repeating pattern. It is possible for a unit cell to possess accidental equalities among its angles or the lengths of its edges that are not demanded by the symmetry. For example, it would be possible for the "N" pattern of Fig. 3.6 to be laid out on a rectangular grid, but it would not be orthorhombic; its unit cells would accidentally have 90° angles, but these angles would not be demanded by the symmetry. Similarly an orthorhombic crystal might accidentally have $a = b$, but it would not thereby become tetragonal if the pattern did not possess 4-fold symmetry. Thus symmetry is primary, and equalities between edges or angles (or special values of angles) are secondary, and are only significant in so far as they arise from the symmetry. It is for this reason that one cannot devise other crystal systems that might at first sight seem to be possible. For example, there can be no "diclinic system" with one interaxial angle of 90°, because there is no symmetry that would require this.

Instructions for viewing stereoscopic drawings

In order to obtain the stereoscopic effect it is necessary that the left eye should attend only to the left-hand member of the pair and the right eye only to the right-hand member. This can be ensured by using a simple stereoscope with two lenses so as to permit the drawing to be viewed in focus at a very short distance from the eyes. However, it is by no means necessary to use such an instrument. If the drawing is brought up very close to the face (100–150 mm) each eye will see both drawings but the image of the right-hand drawing formed by the left eye will be very close to the image of the left-hand drawing formed by the right eye. By adjusting the distance of the drawing from the face these two images can be made to coincide, so that one seems to be looking at three drawings, although they are all out of focus because of the short viewing distance. However, if attention is concentrated on the central image of the three and the other two are ignored, this appearance can be retained as the drawing is gradually taken back to a comfortable distance at which to focus, and the central image will then be seen stereoscopically.

Some people have more difficulty than others in such stereoscopic viewing, but when the trick has been done once it becomes much easier to do it again. If it proves quite impossible and no stereoscope is available, then either member of the pair can be interpreted as an ordinary perspective drawing.

Problems

List all the symmetry elements of:
1. A regular octahedron.
2. A regular tetrahedron.
3. A pyramid on a square base.
4. A match-box.
5. An arrow-head.

CHAPTER 4

Notation for Crystallographic Faces, Forms and Zones

Crystallographic axes

IN order to consider the relationship between crystal faces more systematically their positions are referred to Cartesian axes, x, y, z. These axes are always taken parallel to the edges of a unit cell, so that the a-edge is parallel to the x-axis, and the b-edge to the y-axis, and the c-edge to the z-axis. It follows that the *crystallographic axes* are only orthogonal axes if the crystal belongs to one of the three orthogonal systems (orthorhombic, tetragonal or cubic); otherwise one or more of the interaxial angles will be oblique, and when they are oblique they are denoted α, β, γ corresponding to the equivalent angles between the edges of the unit cell. In perspective drawings involving the crystallographic axes the normal convention is to draw the positive direction of the z-axis vertically upward on the paper, the positive y-axis to the right, and the positive x-axis coming out towards the reader, as in Fig. 4.1.

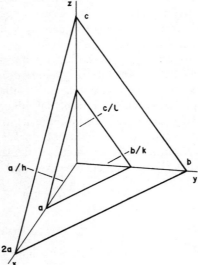

FIG. 4.1. A face that steps two cells in the a-direction, one in the b-direction and one in the c-direction is moved towards the origin until it intercepts $a/1$ on the x-axis, $b/2$ on the y-axis and $c/2$ on the x-axis. The intercepts are marked a/h, b/k, c/l corresponding to a general face (hkl). The specific face shown is (122).

Miller indices

The orientation of a plane can be expressed in terms of the relative values of the intercepts that it makes on a given set of Cartesian axes, and this fact is used to specify the orientation of crystal faces.

As we have seen in Chapter 1, a crystal face is always stepped by whole numbers of unit cells in each of the axial directions, e.g. N_1 in the a-direction, N_2 in the b-direction and N_3 in the c-direction. If we define the x-, y- and z-axes parallel to these directions (and therefore not necessarily at right angles to one another), we can always choose an origin at a corner of a suitable unit cell inside the crystal so that the face intercepts distances of $N_1 a$ on the x-axis, $N_2 b$ on the y-axis and $N_3 c$ on the z-axis, so that the numbers N_1, N_2, N_3 could be used to specify the orientation of the face. However, it turns out to be more convenient to specify it in terms of three numbers derived from these. If we imagine the face to be moved parallel to itself towards the origin, the three intercepts will all decrease in the same proportion (Fig. 4.1), and when they have decreased by a factor N which is the least common multiple of N_1, N_2 and N_3 they will simultaneously be integral fractions of a, b, and c, i.e. $a/(N/N_1)$, $b/(N/N_2)$ and $c/(N/N_3)$. The quantities N/N_1, N/N_2 and N/N_3 are always denoted by h, k and l and it follows from their derivation that h, k and l are whole numbers that do not have a common factor, and can be used to give a unique and unambiguous notation for the face. They are called its *indices* (sometimes its *Miller indices* after their originator) and are written as (hkl) in round brackets, without any commas between them. If the face is parallel to an axis the intercept on that axis is infinite of course, but this is regarded as $a/0$ (or $b/0$ or $c/0$ as the case may be), and the corresponding index is therefore zero. Thus faces parallel to the bc, ca and ab faces of the unit cell are respectively (100), (010) and (001). If a face intercepts the negative end of an axis then the corresponding index is negative, and the negative sign is written as a bar over the top of it; thus the face ($1\bar{2}0$) intercepts a on the positive end of the x-axis, $b/2$ on the negative end of the y-axis, and is parallel to the z-axis. This index notation works equally well regardless of whether the crystallographic axes are orthogonal or oblique.

In the above formulation of the concept of indices, a plane parallel to a crystal face was imagined to be brought gradually in toward the origin until its intercepts were a/h, b/k and c/l. This was done in order to tie the concept firmly to that of the unit cell. But in morphological crystallography we never know the actual values of a, b and c but only their ratios $a:b:c$. Thus the specific position described is quite arbitrary, and in *all* positions of the plane it makes intercepts on the axes whose ratios are $a/h:b/k:c/l$. This means that we never actually have to bother shifting a plane to some particular position; wherever a face (hkl) may lie we can simply extend it until it cuts the axes, and then its intercepts on them are in the ratios $a/h:b/k:c/l$.

Forms and form indices

The symmetry operation corresponding to any symmetry element of a crystal will always either leave a face unchanged or transform it into the position previously occupied by another face. To any face there therefore corresponds a whole set of faces into which it is transformed by the various symmetry operations that form part of the symmetry of the crystal. Such a set of faces, related to one another by the symmetry, is called a *form*. This is a technical term in crystallography with this specific meaning; it does not mean overall shape as it does in everyday life. The number of faces in a form can be very various, depending both on the symmetry of the crystal and the orientation of the face relative to the symmetry elements. Thus a face perpendicular to a rotation axis is not repeated by that axis, but any other face is repeated by it n times if it is an n-fold axis. In a crystal that has no symmetry (like crystals of one of the classes of the triclinic system) every form consists of just one face, since there is no repetition at all. Forms containing only one face also occur in some other crystal classes which have only a single axis of symmetry with no mirror perpendicular to it, if the face concerned is perpendicular to the symmetry axis. By contrast, in the cubic system there are forms containing as many as forty-eight faces.

Forms are of two types, *open forms* and *closed forms*. A closed form is one whose members are capable of totally enclosing a space. For example, the faces on the orthorhombic crystal in Fig. 3.8(b) are all related to one another by the symmetry, and since they are the only faces on the crystal which they bound they are clearly capable of enclosing a space. However, a closed form may co-exist on a crystal with other forms. The crystal in Fig. 3.8(a) possesses faces corresponding to those of Fig. 3.8(b) which are therefore clearly *capable* of enclosing a space although they do not happen to do so in Fig. 3.8(a) because of the presence of other faces as well. An open form, on the other hand, is incapable in any circumstances of enclosing a space on its own. This is obviously true of any form that consists of less than four faces, and it is also true of any form all of whose members are parallel to a single direction. The faces in Fig. 3.8(a) that are not on Fig. 3.8(b) constitute two forms of this kind each containing four faces respectively parallel to the y- and z-axes. In general a crystal may be bounded by either one closed form, or by a combination of several open forms (the minimum number depending on the nature of the forms), or by a combination of two or more closed forms, or by a combination of closed and open forms.

The various possible forms have been given individual names. Some of these such as cube, octahedron and prism have obvious meanings and will be used freely (though it is to be noted that a prism form is generated by rotating a face round a symmetry axis, and it is therefore an open form—the prism does not include bounding faces at the ends as it would do in ordinary parlance). There are, however, many other terms which have no analogues in everyday language. Thus a pair of parallel faces, related by symmetry, is called a *pinacoid*, and this is useful because it occurs so frequently, but much of this complex nomenclature can be avoided by using *form indices*.

The set of faces into which the face (hkl) is repeated is denoted by the form indices $\{hkl\}$ in curly brackets. The individual faces that form part of the form can always be derived from the form indices if the symmetry is known, in ways that will be discussed in Chapter 7. It is sufficient to note here that the other faces in the form $\{hkl\}$ can always be derived from the indices of the face (hkl) either by changing the sign of one or more of the indices, or by permuting the order of some or all of them, or by a combination of both processes. Simple examples are as follows. On a crystal with the maximum symmetry in the orthorhombic system the form $\{100\}$ contains the two faces (100) and $(\bar{1}00)$; in the tetragonal system the form $\{100\}$ contains the four prism faces (100), (010), $(\bar{1}00)$ and $(0\bar{1}0)$; and in the cubic system the form $\{100\}$ contains the six cube faces (100), (010), (001), $(\bar{1}00)$, $(0\bar{1}0)$ and $(00\bar{1})$.

Symbols for zones and zone axes

The direction of a zone axis is always parallel to a possible edge between two possible crystal faces. It is therefore always parallel to an edge of a unit cell, or to a face diagonal or a body diagonal of a unit cell or of a group of unit cells, and therefore also to a line through the origin such that the coordinates x, y, z of any point on it are in the ratio

$$x : y : z = ua : vb : wc$$

where u, v and w are whole numbers (Fig. 4.2). They are the whole numbers of cell edges bounding the figure of which the zone axis is the diagonal, and they can be used as a symbol both of the zone axis and of the zone itself. For this purpose they are distinguished by putting them in square brackets, as $[uvw]$. It is to be noted that they are always called zone symbols, not indices. They differ from face indices in a very important way, in that their definition involves multiplication of the cell edges by the zone symbols, whereas the definition of face indices involves division of the cell edges by the indices.

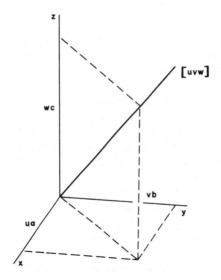

FIG. 4.2. Coordinates of any point on the zone axis $[uvw]$ are in the ratio $ua : vb : wc$.

It follows from the definition of zone symbols that the crystallographic axes x, y, z themselves have the symbols [100], [010] and [001], and they are sometimes referred to in this way.

Calculations using indices and zone symbols

It is demonstrated in books on three-dimensional coordinate geometry that the equation

$$\frac{x}{p} + \frac{y}{q} + \frac{z}{r} = 1$$

represents a plane which cuts intercepts p, q and r off the x-, y- and z-axes respectively, and that a parallel plane through the origin is

$$\frac{x}{p}+\frac{y}{q}+\frac{z}{r}=0.$$

Since the face (hkl) makes intercepts a/h, b/k, c/l on the axes, a plane parallel to it through the origin has the equation

$$\frac{hx}{a}+\frac{ky}{b}+\frac{lz}{c}=0.$$

If this face (hkl) lies in a zone $[uvw]$ it follows that any point on the zone axis, with coordinates ua, vb, wc, lies on this plane and so satisfies the equation

$$\frac{hua}{a}+\frac{kvb}{b}+\frac{lwc}{c}=0,$$

i.e. $$hu+kv+lw=0.$$

It follows that if we have any two faces $(h_1k_1l_1)$ and $(h_2k_2l_2)$ we can find the symbol of the zone which contains them by solving the simultaneous equations

$$h_1u+k_1v+l_1w=0$$
and $$h_2u+k_2v+l_2w=0.$$

It is possible to find the three symbols u, v, w from only two equations, because it is really only their ratios that we have to find; our knowledge that they are integers without a common factor then enables us to find the three of them unambiguously. The solution is

$$u=k_1l_2-k_2l_1,\quad v=l_1h_2-l_2h_1,\quad w=h_1k_2-h_2k_1.$$

A convenient mnemonic for finding it is to write down the indices of the faces twice on consecutive lines,

$$\left.\begin{array}{c}h_1\\h_2\end{array}\right)\begin{array}{c}k_1\\k_2\end{array}\begin{array}{c}l_1\\l_2\end{array}\begin{array}{c}h_1\\h_2\end{array}\begin{array}{c}k_1\\k_2\end{array}\left(\begin{array}{c}l_1\\l_2\end{array}\right.$$

to ignore the first and last indices, and to cross multiply the remainder as shown, upward going arrows giving negative products.

The såme process obviously operates in reverse; given two zones whose symbols $[u_1v_1w_1]$, $[u_2v_2w_2]$ are known, the face which they have in common is given by the solution of

$$hu_1+kv_1+lw_1=0$$
and $$hu_2+kv_2+lw_2=0,$$

that is

$$h=v_1w_2-v_2w_1,\quad k=w_1u_2-w_2u_1,\quad l=u_1v_2-u_2v_1.$$

This can be obtained by the same mnemonic, writing each zone symbol twice.

Either of these processes may sometimes lead to a set of indices or zone symbols containing a common factor, and in this event they should always be divided through by the factor. If the members of the set are all negative this can be regarded as equivalent to a common factor of -1, and all the signs can be changed. Obviously the zone axis $[\bar{u}\bar{v}\bar{w}]$ is simply the same line as $[uvw]$ on the other side of the origin, so changing the sign of a whole set of zone symbols makes no difference. The faces (hkl) and $(\bar{h}\bar{k}\bar{l})$ are of course different from one another and opposite to one another, but two zones always possess, a pair of opposite faces in common, and which of the two emerges from the calculation depends merely on which of the two sets of zone symbols is written down first.

Another useful relationship between indices is the fact that if a face (hkl) lies in the same zone as two other faces $(h_1k_1l_1)$ and $(h_2k_2l_2)$ then the indices of the first face can be expressed as the sums of integral multiples of the corresponding indices of the other two:

$$h = mh_1 + nh_2, \quad k = mk_1 + nk_2, \quad l = ml_1 + nl_2.$$

This provides a convenient way of checking whether a given face is in the same zone as two others without calculating any zone symbols. It is often very easy to see whether such a relationship holds because indices are often such very small numbers, like 0 and 1. Thus, from this relationship, the face (111) obviously lies in the zone containing (110) and (001), and also in the zone containing (100) and (011). Note that such zonal relationships are universal, and independent of the crystal system.

Reformulation of the process of finding axial ratios

In Chapter 1 we approached the problem of finding axial ratios in an intuitive but rather cumbersome way in terms of postulations about the orientations of the unit cells with respect to the faces, and how the latter were "stepped" in terms of the unit cells. It is now possible to specify the procedure much more elegantly in terms of symmetry and indices. The crystal in Fig. 1.3 has 2-fold rotation axes perpendicular to the faces A, B and C. Thus it is orthorhombic, and these directions are therefore taken as the crystallographic axes x, y, z. The faces A, B, C are accordingly (100), (010) and (001). The face P lies in a zone containing (100) and (001) and may therefore be legitimately assigned any set of indices that is a linear combination of these, and most simply (101). If the angle between (100) and (101) is measured and found to be θ, then the xz section of the crystal can be drawn as in Fig. 4.3. The (101) face is extended to cut the x- and y-axes as shown, and the $a:c$ ratio is

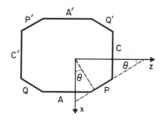

FIG. 4.3. The xz-section of the crystal of Fig. 1.3.

evidently the ratio of the two intercepts, which is $\tan \theta$. Similar considerations apply to the vertical yz section. The face S may be assigned the indices (011) and its intercepts give the ratio $b:c$. The intercepts of the faces F and G on the x- and y-axes then enable us to find their indices, (120) and (110) respectively. The indices of other faces can then be found either from their intercepts or from zonal relationships. For example, U is in a zone between S (011) and G (110) so its indices are a linear combination of these. It is also in a zone between B (010) and P (101), and a few trials quickly show that, in order to be a linear combination of these as well, its indices must be (121).

Alternatively we could set out this indexing procedure more formally. The zone containing (010) and (101) has zone symbols given by the cross-multiplication process

$$\begin{pmatrix} 0 \\ 1 \end{pmatrix} \begin{matrix} 1 & 0 & 0 & 1 \\ 0 & 1 & 1 & 0 \end{matrix} \begin{pmatrix} 0 \\ 1 \end{pmatrix},$$

i.e. $[10\bar{1}]$. Similarly the zone containing (001) and (110) is $[\bar{1}1\bar{1}]$. The face U that lies in both these zones therefore has its indices given by the cross-multiplication

$$\begin{pmatrix} 1 \\ \bar{1} \end{pmatrix} \begin{matrix} 0 & \bar{1} & 1 & 0 \\ 1 & \bar{1} & \bar{1} & 1 \end{matrix} \begin{pmatrix} \bar{1} \\ \bar{1} \end{pmatrix},$$

i.e. (121).

Problems

1. What are the indices of the faces A, B, C, P, Q, G, F, S and T on the crystal shown in Fig. 1.3(b), assuming that its unit cells are as shown in Fig. 1.3(a)?
2. What are the symbols of the zones containing (a) EDAGFB; (b) TPB, (c) GS? Hence find the indices of face U which belongs to both the last two zones.
3. What are the indices of the same faces and the symbols of the same zones in terms of the unit cell that resulted from the assumptions made in question 1 on Chapter 2 (p. 16)?
4. What are the indices of the same faces, and the symbols of the same zones, on the crystal shown in Fig. 1.5(b), in terms of the unit cell that resulted from the assumptions made in question 4 on Chapter 2 (p. 16)?
5. A crystal of the cubic system is in the form of a regular octahedron. Its crystallographic axes lie along the 4-fold symmetry axes. What are the indices of the eight faces and the form indices of the octahedral form? What zones are present on the crystal?

CHAPTER 5

Use of the Stereographic Projection

Well-formed crystals

WHEN a crystal grows in ideal conditions every face in a given form has an equal chance to grow, and when the crystal has finished growing faces related by the symmetry are therefore of the same size. Such a crystal is said to be *well-formed*, and it is easy by looking at such a crystal to see what symmetry elements it possesses. All the diagrams presented so far have depicted well-formed crystals. However, real crystals are rarely well-formed because there are usually influences during their growth which lead to inequalities in the opportunities for growth in different directions. An obvious example is shown in Fig. 5.1. A small, initially well-formed orthorhombic crystal having perpendicular mirror planes is shown in cross-section at (a) lying with one of its faces against another object, which inhibits its growth on that side. After a growth layer of constant thickness has been laid down on the accessible faces of the crystal as at (b) its shape no longer conforms to the symmetry – two of the four symmetrically related faces are smaller than the other two – and mere inspection of the shape of such a crystal will probably fail to reveal its true crystallographic symmetry. However, the directions of the faces are unchanged, and these are capable of revealing the true symmetry providing that information about them can be isolated from the more evident, but confusing, information about the relative sizes of the faces.

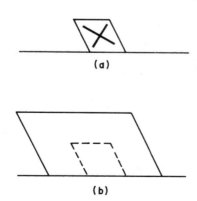

FIG. 5.1. (a) A well-formed crystal (with mirror planes on the diagonals) is impeded from growing at one side. (b) It grows into a larger crystal whose shape does not exemplify the true symmetry.

The spherical projection of a crystal

For this purpose, imagine a sphere (substantially larger than the crystal) centred on some arbitrary point inside the crystal (Fig. 5.2). From this point construct normals to the faces, and extend these normals to cut the sphere. Then the points on the sphere where the normals emerge represent the faces of the crystal in terms of direction, entirely uninfluenced by their relative sizes, and the symmetry of the arrangement of these points on the surface of the sphere will reveal the true symmetry of the crystal whether or not it be well-formed. Such a representation is called a *spherical projection* of the crystal.

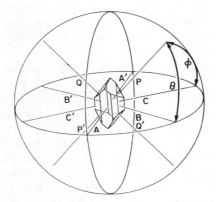

FIG. 5.2. The spherical projection of a crystal.

The point in which a normal to a face cuts the spherical projection is called the *pole* of that face, and it is to be noted that the poles of all the faces that lie in a zone always lie along a great circle of the sphere. This is because the faces in a zone are all parallel to their zone axis, to which their normals are therefore all perpendicular; these normals therefore lie in a plane through the centre of the sphere (they were all drawn from that point), cutting the sphere along a great circle. Since interfacial angles are defined as being the angles between the normals to faces, the angular distance along the great circle joining the poles of two faces is equal to the angle between the faces. In particular if a face A' is opposite to a face A, then the pole of A' lies on the sphere at the "antipodes" of the pole of A. It is frequently useful to use such geographical terminology in discussing the spherical projection of a crystal, especially the idea of the equator and the north and south poles of the sphere.

In the example shown in Fig. 5.2 the crystal has a prominently developed zone of faces, and since the axis of this zone is almost certain to turn out to be a convenient crystallographic axis, it is appropriate to orient the crystal as shown with the prominent zone parallel to the imagined north–south axis of the sphere. The poles of the faces in the prominent zone therefore lie on the equator. When the angles between them have been measured, the pole of one of them is arbitrarily placed on the zero meridian, and then the others can all be positioned at the appropriate angles of longitude round the equator. The pole of any other face such as P can be positioned on the sphere if the interfacial angles are measured from P to two of the faces already plotted, say B and C. If the angle between P and B is θ, then the pole of P must lie on a small circle of radius θ around the pole of B. This is a circle of points on the sphere which are all an angular distance θ away from the pole of B as measured along the great circles joining them to it. Similarly, if the angle between P and C is ϕ then the pole of P must lie on a small circle of radius ϕ around the pole of B. Thus the pole of P is at the intersection of these two circles on the upper half of the sphere.

The stereographic projection and its properties

Although it is possible to construct an actual spherical projection of a crystal, it is not really very convenient to do so. Just as in geography a map is for most purposes easier to make than a globe, so in crystallography it is usual to map the sphere on to a plane, and the most useful projection of the sphere for this purpose is the *stereographic projection*.

The stereographic projection is defined as shown in Fig. 5.3. Any point on the surface of the sphere is projected on to the equatorial plane by joining it to the south pole, the intersection of this join with the equatorial plane being the projection of the point in question. This is straightforward for points on the northern hemisphere, which all plot within the equatorial circle; for points on the southern hemisphere it is necessary to join them to the south pole and then produce these joins backwards. Such points project to points outside the equatorial circle. The north pole obviously projects to the centre of the projection, but the south pole cannot be represented, because as a point on the sphere tends towards the south pole its projection tends to an infinite distance from the centre of the projection. The equator of the sphere obviously projects to itself, the equatorial circle on the projection plane, and this is called the *primitive* of the projection.

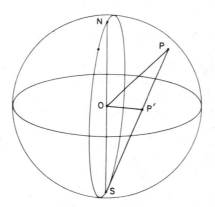

FIG. 5.3. The point P' is the stereographic projection of the point P on the sphere.

The features of the stereographic projection that make it so useful in crystallography are that *all circles on the sphere project into circles*, and that it is relatively easy to recognise the symmetry of a crystal from a stereographic projection of the poles of its faces. However, in order to acquire the skill of doing this it is necessary to get some feeling for the way that the projection works.

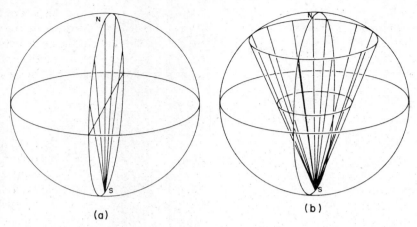

FIG. 5.4. (a) A meridian on the sphere projects into a diameter of the primitive.
(b) A circle of latitude projects into a circle concentric with the primitive.

It is evident from Fig. 5.4 that a meridian of the sphere projects to a diameter of the primitive, and a circle of latitude projects to a circle whose centre is the centre of the primitive. To find the radius of such a circle, consider the vertical section of the sphere in Fig. 5.5(a). Mark off points at angles of 30°, 60° and 120° from the north pole. Then the distances OA, OB, OC are the radii of the projections of the circles of latitude at 60°N, 30°N and 30°S. These projections are drawn in Fig. 5.5(b), and diameters of the primitive are also drawn in at every 30°. Thus Fig. 5.5(b) is a stereographic projection of the lines of latitude and longitude* at every 30° interval on a globe, as far as 30°S. Note that distances on the projection that correspond to equal angles on the sphere increase with distance from the centre.

A similar procedure may be adopted to solve the very useful exercise of finding the stereographic projection of a small circle of given angular radius, θ, around some particular point, P, on the sphere. Figure 5.6(a) shows a vertical section of the sphere through P; A and B are at latitudes θ above and below that of P; and p, a, b are the stereographic projections of P, A, B along a radius of the primitive. On transferring the positions of p, a, b to a drawing of the projection itself we obtain Fig. 5.6(b).

Thus a and b lie on the projection of the small circle of radius θ around P, and the projection of this small circle must also be a circle (since all circles are projected into circles by the stereographic projection); however, the centre of this circle is clearly not p itself, but must be a point mid-way between a and b, and the radius of the circle is a half of the length ab.

Another procedure that is often required is to find the stereographic projection of the great circle through two points P and Q on the sphere. This will clearly be a circle through p and q, the projections of P and Q, but to define this circle we need to know the position of a third point that lies on it. Since every great circle through a point P passes through the

---

* The apparent anomaly that the meridians are represented by straight lines, although it has been stated that every circle projects to a circle, is clarified on p. 44.

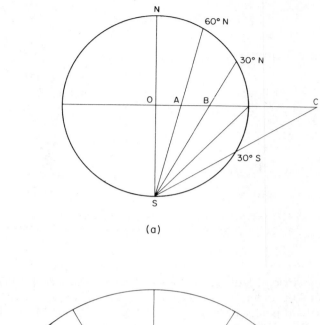

(a)

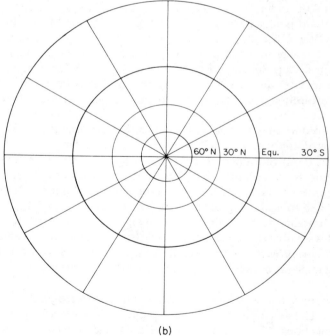

(b)

Fig. 5.5. (a) Construction to find the radii of the stereographic projection of circles of latitude at 60° N, 30° N and 30° S. (b) The corresponding stereographic projection, including the meridians at 30° intervals.

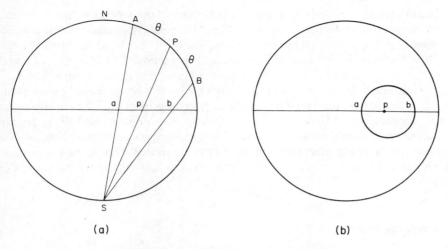

FIG. 5.6. (a) Vertical section of the sphere through P. (b) Stereographic projection of a small circle of radius θ around P.

point P' opposite to it (its "antipodes"), the problem is solved by finding p', the projection of P'. Figure 5.7(a) shows a vertical section of the sphere through P, which therefore also contains P'. If P and P' are joined to the south pole the positions of p and p' are found along the diameter of the primitive through p. On transferring these positions to the stereographic projection in Fig. 5.7(b) the construction can be completed; the centre of the required circle must be equidistant from p and p' and so lies on the perpendicular bisector of the line pp'. Similarly it must lie on the perpendicular bisector of pq, and so can be found at the intersection of these two lines at c. The circle itself can then be drawn.

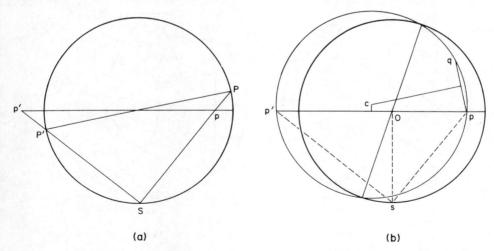

FIG. 5.7. (a) Preliminary construction to find the projection p' of P', the antipodes of P. (b) Construction of the stereographic projection pqp' of the great circle through P, Q and P'. Broken lines indicate how (a) could have been performed on (b) itself.

Actually the construction of the preliminary diagram of the vertical section of the sphere in Fig. 5.7(a) was not really necessary in practice. All the necessary construction lines could have been drawn on the stereographic projection itself, as shown in broken lines on Fig. 5.7(b). To find p' it suffices to join Op, draw Os perpendicular to Op, join ps, and then construct a line sp' perpendicular to ps to meet pO produced in the required point p'. All these lines are equivalent to those in Fig. 5.7(a) if the latter is imagined to be rotated about Op until it lies flat on the plane of the stereographic projection. The constructions of Fig. 5.6(a) could similarly have been done on the projection itself, and all stereographic constructions that should notionally be done on a vertical section of the sphere can always be carried out on the projection itself, as it is being drawn. It is to be noted that the projection of any great circle cuts the primitive in two points that lie at opposite ends of a diameter.

The stereographic net

One of our main reasons for using the stereographic projection is to enable us to recognise the symmetry of arrangements of poles of crystal faces represented on it. If we are looking for a possible axis of symmetry then we have to be able to envisage the effect of a rotation of the crystal about that axis. Clearly Fig. 5.5(b) helps us to visualise such a rotation if it happens to be about the NS axis; during such a rotation every point will move along its own line of latitude, and angular distances around the lines of latitude are marked off by the meridians. If the rotation were about some other direction it would be very helpful to have a stereographic projection of the lines of latitude and longitude on a globe whose axis was tilted as in Fig. 5.8(a); projected not on its own equatorial plane, but on its horizontal xy-section by joining each point on the sphere to its lowest point and using that

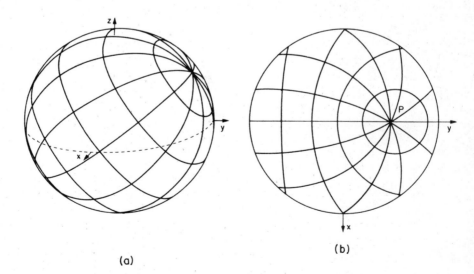

FIG. 5.8. (a) Titled sphere with lines of latitude and longitude.
(b) Stereographic projection of (a) on the xy-plane.

as a new "south pole". Such a projection can be constructed using only the methods we have discussed so far, and is shown in Fig. 5.8(b) for the upper half of the sphere only – the part which projects inside the primitive. Again, during rotation about an axis that emerges at P in the projection a point will move (in projection) along its own lines of latitude, and angular distances around this are marked off by the meridians.

Although Fig. 5.8(b) is very helpful in indicating how we are to regard rotations about an inclined axis, it would obviously be impracticable to have such charts for all possible angles of inclination of the axis. However, there is one chart, the *Wulff net* shown in Fig. 5.9, which is so useful that it is often referred to simply as *the* stereographic net. It

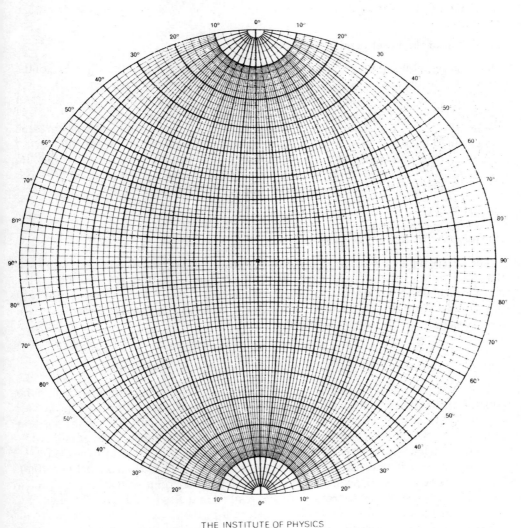

THE INSTITUTE OF PHYSICS

FIG. 5.9. The 5-inch diameter Wulff net.

corresponds to Fig. 5.8(b) except that the axis of the globe has been tipped right over on to the x-axis. The Wulff net therefore gives a direct indication of how a point will move on the projection when rotated about such an axis, each point of course still moving along its own line of latitude (a nearly horizontal arc on Fig. 5.9), angular distance around this being marked off by the meridians. Wulff nets are published in various sizes, the most convenient being of 5 inch diameter with all the circles at 2° intervals.

From Fig. 5.9 it is now possible to explain the fact that was noted on p. 39 that the meridians in Fig. 5.5 projected into straight lines. Figure 5.9 shows that the nearer the projection of a great circle is to the centre of the primitive the larger is its radius, and the diameters of the primitive are to be regarded as parts of circles of infinite radius.

Projection of the lower hemisphere to the north pole

In both Figs. 5.8 and 5.9 the projection has been confined to that of the upper half of the sphere, which projects within the primitive, since this provides a convenient stopping place: once one goes outside the primitive there is no obvious stopping place short of infinity, which is very inconvenient. In stereographic projections of crystals the inconvenience of a projection that goes off to infinity is usually avoided by plotting the projection of the lower part of the crystal to the topmost point of the sphere. This part of the crystal then also comes within the primitive, and features of the top and bottom are distinguished by using appropriately different symbols: for example, a point on the top is represented by a · or an ×, and a point on the bottom is represented by a little circle ○.

Plotting the stereogram of a crystal

The stereographic projection of the poles of the faces of a crystal is commonly known as the stereogram of the crystal. It may be constructed directly from information on the interfacial angles, either by using the fundamental constructions discussed above, or with the help of a Wulff net. The method will be illustrated by reference to the crystal shown in Fig. 5.2. The angles between the faces are:

$$AB = 56°, BC = 68°, CA' = 56°, A'B' = 56°, B'C' = 68°, C'A = 56°,$$
$$BP = CP = B'Q = C'Q = BQ' = CQ' = B'P' = C'P' = 43°.$$

In Fig. 5.2 we chose to represent the poles of the faces in the vertical prism zone around the equator of the sphere, and correspondingly we mark them on the primitive (Fig. 5.10(a)) with the help of a protractor, starting from an arbitrary position for A. To find the position of P, we have to construct around B and C the (projections of) small circles of radius 43°; the point where these intersect within the primitive is the projection of the pole of P.

If the circles are to be constructed from scratch the procedure is as shown in Fig. 5.10(a). To find the circle about B, points d, e are marked off on the primitive at 43° from B, and a point s 90° from B. The intersections d' and e' of sd and se with OB give the ends of a diameter of the required circle. (This procedure uses the device described on p. 42 of temporarily using the primitive, when it suits us, as the equivalent of the vertical section of the sphere.) The small circle can now be drawn, centred at the mid-point of d'e'. The small circle of radius 43° about C' (note the conventional shortened nomenclature whereby we speak of the stereographic projection as though it were the sphere itself) is constructed similarly, though the details of the process are not shown in the figure, and the two small circles intersect in P, the pole of that face. From the symmetrical form of the angular data, Q can be inserted at a point along PO produced so that OQ = OP. The position of Q' is obviously at the second intersection of the two small circles, outside the primitive. However, it is much more convenient to replot P' and Q' by projecting their poles to the north instead of to the south pole, so as to bring them within the primitive. From the angles, they then obviously lie at the same places in the projection as Q and P respectively, and the conventional finished projection is that of Fig. 5.10(b). Prominent zones have been indicated by great circles and the symmetry elements have also been marked in, using heavy lines along great circles where mirror planes intersect the sphere. The symbols of symmetry axes are added where these emerge from the sphere. The presence of these symmetry elements is deduced from the positions of the poles on the stereogram. It is the full (maximum) symmetry of the orthorhombic system.

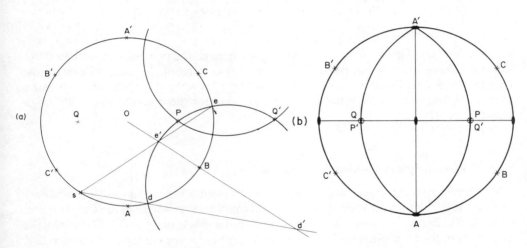

FIG. 5.10. (a) Construction of the stereogram of the crystal in Fig. 5.2. (b) The finished stereogram with symmetry elements inserted.

If the stereogram is to be constructed with the help of the Wulff net it is best to draw the stereogram on tracing paper. The primitive is drawn first and superimposed on the Wulff net. The prism zone can then be marked in directly using the angular graduations of the net itself. The net is then rotated with respect to the drawing to make one of its "poles" coincident with B, and an arc of a small circle of radius 43° traced on the stereogram (Fig. 5.11(a)). The net is then rotated to make the same "pole" coincide with C and a similar arc is traced and the intersection of the two arcs defines the position of the face P. The stereogram can then be completed as before.

The Wulff net also provides a very convenient way of finding from the stereogram any interfacial angle that has not been measured. All that is required is to rotate the stereogram

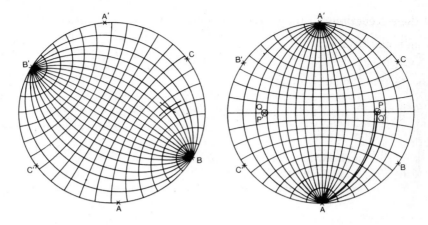

FIG. 5.11. (a) Construction of the stereogram using the Wulff net.
(b) The finished stereogram with superimposed Wulff net.

over the net until a great circle on the latter goes through the two poles concerned. Then the intervening graduations provided by the intersections of the small circles can be counted to give the interfacial angle. Thus on Fig. 5.11(b) the orientation is such that a great circle passes through A and P, and the graduations show that there is a 90° angle between them.

Use of the stereogram for visualising symmetry

If a stereogram of a crystal is plotted in such an orientation that an axis of symmetry plots at the centre then the stereogram possesses rotational symmetry of the same order as that axis. Similarly if a plane of symmetry intersects the sphere in a great circle that plots as a diameter of the primitive then the stereogram possesses mirror symmetry about that line. Mirror symmetry in the equatorial plane of the sphere is equally obvious since it causes points in the northern hemisphere projected to the S-pole all to coincide with points on the southern hemisphere projected to the N-pole.

If a symmetry element plots in a more general position then its effect is not quite so obvious, but it may still be recognised with a little practice. A mirror plane must always plot along a great circle; if this is considered to lie along one of the great circles of the Wulff net (Fig. 5.9) then any pair of points related by the mirror will lie at equal numbers of divisions to either side of it along one of the small circles. The effect of an axis of symmetry can be understood by reference to Fig. 5.8(b). If it plots at a point such as P, then any set of points related to one another by that axis will plot around one of the small circles round P; their intervals round such a small circle will be unequal in length but equal in terms of the divisions of such a small circle by great circles intersecting in P, those shown in Fig. 5.8(b) being at 30° intervals. Of course, charts of this form are not available for general locations of P, but the effect is not difficult to visualise in a qualitative way.

Use of the stereogram of a crystal to find axial ratios

Having found from its stereogram that the crystal of Fig. 5.2 is orthorhombic we can add to the stereogram appropriate directions of the crystallographic axes, and proceed to index the faces (Fig. 5.12). The zone ABCA'B'C' is parallel to the z-axis, so the l index of all these faces is 0. A and A' are indexed as (100) and ($\bar{1}$00), and B is assigned the indices (110), so the remaining faces, C, B', C' are ($\bar{1}$10), ($\bar{1}\bar{1}$0) and (1$\bar{1}$0) respectively. Although there is no (001) face on the crystal, it is a possible face and we know that if it were present it would lie on the stereogram at the centre of the primitive. Similarly (010) would lie on the primitive at the point of emergence of the y-axis half-way between (110) and ($\bar{1}$10). Therefore the face P lies on a zone between (010) and (001); its indices are therefore some linear combination of these, (0kl) with k and l indeterminate, but we are entitled to set k and l equal to unity and index P as (011), thereby implicitly defining the choice of unit cell.

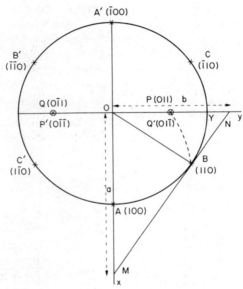

FIG. 5.12. Use of the stereogram of Fig. 5.10(b) to find the axial ratio
$a : c$.

The ratio $a : b$ can now be found very easily. If we draw the tangent to the primitive at the point B, this represents the direction of the face B itself where it cuts the equatorial plane. Thus the intercepts of this tangent on the x- and y-axes are the intercepts of the face (110) on these axes, and so by definition are in the ratio $a : b$. We can either measure their lengths on the drawing and divide one by the other, or else simply calculate their ratio trigonometrically from the right-angled triangle OMN, using the fact that the angle ONM is equal to the angle MOB and this is the same thing as the interfacial angle AB.

The ratio $b : c$ can be found from the position of (011). Using the Wulff net orientated with p at the point of emergence of the y-axis we can measure the angle between "(001)" and (011) from the number of graduations between O and P, and in the usual way we can then show that c/b is the tangent of this angle.

Accurate calculations from the stereogram

The foregoing exercise has demonstrated that the stereogram has potentialities beyond the mere recording of the interfacial angles of a crystal and visualisation of its symmetry; it also suggests ways of calculating axial ratios, and of obtaining graphically the values of angles that help in such calculations. However, angles obtained by such graphical methods are not very accurate. Reading from a 5-inch Wulff net is not accurate to much better than 1°, and the use of the fundamental constructions on a larger scale would only improve this accuracy to a modest extent, whereas the interfacial angles measured on an optical goniometer may be accurate to a minute of arc. In order to make use of the accuracy of such measurements it is therefore necessary to be able to calculate one angle from another. This is done by use of spherical trigonometry.

Spherical trigonometry permits the solution of spherical triangles whose sides consist of arcs of great circles on a sphere. The sides of such a triangle ABC (Fig. 5.13) are denoted a, b, c and are expressed not as lengths but as the angle subtended by the arc at the centre of the sphere; and the angles at the vertices are defined as the angles between the planes whose intersections with the sphere define the great circles involved. Thus the triangle has three angles A, B, C, and three sides a, b, c also expressed in angular measure; all these are actually angles in the three-dimensional space inside the sphere, and by drawing various ordinary plane triangles involving these angles it is possible to work out relationships between them. These relationships can then be used to calculate unknown sides or angles of a spherical triangle if a sufficient number of the others are known.

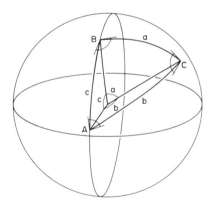

FIG. 5.13. A spherical triangle ABC showing the relation of its sides and angles within and on the sphere.

The two most important formulae for solving spherical triangles are the sine rule and the cosine rule. The former strongly resembles the ordinary sine rule for plane triangles. It is:

$$\frac{\sin a}{\sin A} = \frac{\sin b}{\sin B} = \frac{\sin c}{\sin C}.$$

Use of the Stereographic Projection

The cosine rule for spherical triangles is rather different from that for plane triangles. It is

$$\cos c = \cos a \cos b + \sin a \sin b \cos C.$$

In using these relationships two differences between spherical and plane triangles have to be borne in mind. The first is that the sum of the three angles of a spherical triangle, $A+B+C$, is *not* 180°; it is always more than 180°, and the excess over 180° increases with the size of the triangle. This fact seems initially surprising, but is clarified by Fig. 5.14 in which the triangle NPQ, bounded by two meridians and the equator, obviously has $N+P+Q$ greater than 180°, since P and Q are each equal to 90°. The second difference is that there are no such things as similar spherical triangles. This is essentially because one cannot have parallel great circles. It is closely related to the previous point about the sum of the angles. In Fig. 5.14 it is evident that triangles like NPQ and NP′Q′ are not similar in that they have different angles at P and P′ and Q and Q′. It may be seen that these angles diminish with the size of the triangle since the meridians are perpendicular to the circles of latitude (broken line) and the great circle P′Q′ lies above this circle. Since the angle at N remains the same it follows that the sum of the three angles of the triangle diminishes with its size. Because of this property one must always use the sine rule or cosine rule to find the third angle of a spherical triangle even when the first two are known.

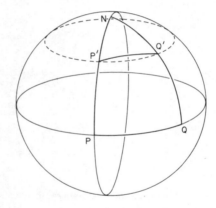

FIG. 5.14. The spherical triangle NPQ has right angles at both P and Q, but NP′Q′ does not have any right angles.

We can now consider how to calculate the angle OP that we previously had to measure with the Wulff net on Fig. 5.12. Triangles on the stereogram that are formed by the projections of great circles represent spherical triangles and can be regarded as such, since we measure their sides in terms of angular measure on the sphere. If we denote by Y the intersection of the y-axis with the primitive on Fig. 5.12, then BPY is a spherical triangle in which we know

$$BP = 43°,$$
$$BY = \tfrac{1}{2}BC = 34°,$$

and the angle $Y = 90°$.

Then, since $\cos Y = 0$, the cosine rule simplifies to

$$\cos BP = \cos BY \cos YP$$

so that $\cos YP = \cos 43°/\cos 34°$

and $YP = 28.1°$,

which enables us to calculate the ratio $b:c$.

Although the sine rule and cosine rule would suffice to solve any spherical triangle with which we are faced, it is often more convenient to use special forms to deal with right-angled spherical triangles, just as we do in dealing with plane right-angled triangles. For this purpose one uses Napier's Rule. A right-angled triangle contains five variables, the three sides and the two other angles. These are numbered cyclically (1), (2), (3), (4), (5) starting from a side adjacent to the right angle, as in Fig. 5.15(a). They are then associated with the five regions of the diagram in Fig. 5.15(b). Here the horizontal line on the right represents the position of the right angle, and the five quantities indicated are called the five "parts" of the triangle. Napier's Rule states that:

sine of any part = product of cosines of opposite parts

and   sine of any part = product of tangents of adjacent parts.

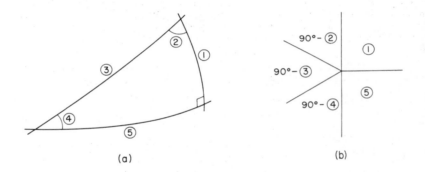

Fig. 5.15. (a) A right-angled spherical triangle. (b) Napier's diagram of the five parts.

(The alliteration helps in remembering these formulae.)

If any two parts are known it is then possible to find any other. For example if we wish to calculate (1) this can be done from any of the six relationships.

$$\begin{aligned}
\sin(1) &= \cos(90° - (3))\cos(90° - (4)), \\
\sin(1) &= \tan(90° - (2))\tan(5), \\
\sin(90° - (3)) &= \cos(1)\cos(5), \\
\sin(90° - (4)) &= \cos(1)\cos(90° - (2)), \\
\sin(90° - (2)) &= \tan(1)\tan(90° - (3)), \\
\sin(5) &= \tan(1)\tan(90° - (4)),
\end{aligned}$$

according to which two of the other four parts we know. When using Napier's Rule it is

Use of the Stereographic Projection 51

important always to draw the equivalent of Fig. 5.15(b) with the actual values of the angles inserted, in order to avoid confusion.

In addition to the right-angled spherical triangles discussed above, a second kind of right-angled spherical triangle is also possible, namely one in which one of the sides is a right angle. Such triangles can be treated by a modification of Napier's Rule in which each of the variables (1) to (5) is replaced by its supplement (i.e. $180° - (1)$, etc.) before being inserted in the Napier diagram to give the corresponding "part" of the triangle.

Further examples of the use of spherical trigonometry in calculating unit cell parameters will be found in the next chapter.

Problems

1. Draw on tracing paper a circle of the same size as the Wulff net in Fig. 5.9. Superimpose it on the net and draw a sterographic projection of the earth showing the positions of London (51°N 0°W), Mexico City 19°N 99°W), Delhi (29°N 77°E) and Sydney (34°S 151°E), plotting the last within the primitive by projection to the N-pole. Use the net to find the great circle distances between the six possible pairs of cities, assuming that the radius of the earth is 6350 km. Remember that you can rotate the tracing paper over the net to make the measurements that you want.
2. The ill-formed crystal shown in the diagrams has the following interfacial angles: between adjacent faces in each of the zones a c e g a, a j e m a, and c j g m c every angle is 90°; between adjacent faces in each of the zones s j t u m v s, w j x y m z w, z a s x e u z, w a v y e t w, the angles are 55°, 55°, 70°, 55°, 55°, 70°, respectively. Plot a stereogram of the crystal on tracing paper with the help of the Wulff net in Fig. 5.9. Hence determine the system to which the crystal belongs, and draw in the zones a s x e u ż a, a w t e y v a, etc. Index the faces on the basis of the symmetry and the given angles.

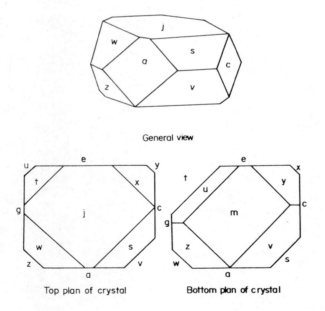

3. An aircraft flying on a great circle route passes 13°51′ due north of Quito (0°N 99°W) and then passes 17° 37′ due west of it. Calculate by spherical trigonometry how far the aircraft has travelled between these two points and the bearing of its flight path at each of them. (Radius of the earth is 6350 km.)

4. The (111) face of an orthorhombic crystal makes angles of 65° 33' with (100) and 51° 21' with (010). Draw a rough sterogram (not to scale) showing the positions of these three faces and also of (001). Draw the zones through (001) and (111), and (100) and (111) and hence mark the positions of (110) and (011). Use the spherical triangles on the diagram to calculate the angles (100)^(110) and (010)^(011). Hence find $a:b:c$.

CHAPTER 6

Morphology of the Seven Crystal Systems

THIS chapter deals systematically with the various kinds of forms that can be developed in each of the seven crystal systems, although attention is confined to the *holosymmetric crystal class* in each system—that is to crystals possessing the maximum symmetry of the system. These suffice to illustrate for each system the principles of indexing, the relationship of the axes to the symmetry, and appropriate methods of determining axial ratios and interaxial angles, all of which are applicable to all the classes in a given system. Some differences are produced in the forms when the symmetry is lowered to that of one or another of the lower symmetry classes in each system, but a general discussion of the forms in all the thirty-two crystal classes is postponed until Chapter 7.

Discussion of the forms in each system is assisted by considering their spherical projections, represented of course by their stereographic projections. Stereograms are used in this connection to represent purely qualitatively the orientations of faces relative to symmetry elements, e.g. whether they are or are not perpendicular to an axis or a mirror plane. Thus the pole of a face may be put in some arbitrary position on the stereogram, and the way in which it is repeated by the symmetry can then be considered and developed on the stereogram in order to find the nature of the form to which it gives rise. Stereograms of this sort, which show the symmetry elements and the poles of the faces of one or more forms of some purely hypothetical crystal are called *sketch stereograms*. Their construction is greatly helped by considering the way in which the surface of the sphere is divided up by the mirror planes of a particular system into areas of equivalent shape such that any one of them is repeated by the symmetry to cover the whole sphere. These areas are often spherical triangles and are then known as *representative triangles*. The way that they are used will be seen as the discussion develops.

The order in which the systems are discussed is the same as in Chapter 3: that is, the orthogonal systems in order of increasing symmetry; then the monoclinic and triclinic in order of increasing complication due to their inclined axes; and finally the hexagonal and trigonal systems which are closely related to one another and have some distinctive features.

The orthorhombic system (holosymmetric class)

In this class there are three mutually perpendicular mirror planes with three mutually perpendicular 2-fold rotation axes along their lines of intersection. These, together with the crystallographic axes in their conventional orientation, are shown stereographically in

54   Crystallography for Earth Science Students

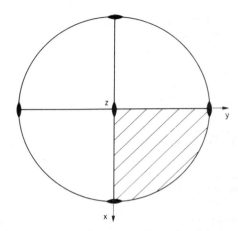

Fig. 6.1. Orthorhombic symmetry represented on the stereogram, with a representative triangle shaded.

Fig. 6.1. The representative triangle is an octant, and the positive octant is shaded; its vertices are the representations of 2-fold rotation axes. If the pole of a face is inserted somewhere within the representative triangle it will be repeated eight times by the symmetry as shown in Fig. 6.2(a). Such a face has no special relationship to any symmetry element; it intersects all three crystallographic axes and has indices of entirely general type (hkl), where h, k and l may have any non-zero integral values. The form $\{hkl\}$ is therefore said to be the *general form*, and it contains eight faces forming a *bipyramid* as illustrated in Fig. 6.2(b). It can equally well be regarded as a bipyramid in three orientations, and each axial plane intersects it in a rhombus. The indices of the faces in this form are obtained by allowing each index to take a +ve or −ve sign independently of the others, i.e. (hkl), ($\bar{h}kl$), ($\bar{h}\bar{k}l$), ($hk\bar{l}$), ($\bar{h}k\bar{l}$), ($\bar{h}\bar{k}\bar{l}$), ($h\bar{k}\bar{l}$), ($\bar{h}\bar{k}\bar{l}$). There are of course an indefinite number of actual

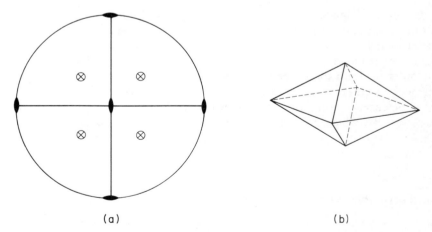

Fig. 6.2. General form $\{hkl\}$ of the holosymmetric class of the orthorhombic system; (a) sketch stereogram; (b) the faces of the form, a bipyramid.

forms of this type depending on the numerical values of the indices, such as {111}, {211}, {121}, etc. These are simply different examples of the same general form.

If the pole that is put into the stereogram is moved on to one of the edges of the representative triangle, then the corresponding face is perpendicular to a mirror plane that no longer repeats it. This therefore gives a *special form*, so called because its faces are in a specific relationship to the symmetry elements. If the pole is placed on the primitive as in Fig. 6.3 then the form is evidently {hk0} and contains four faces in a *prism form* parallel to the z-axis as shown. If the pole had been placed on one of the other sides of the representative triangle then one of two other possible four-faced prism forms, {0kl} or {h0l}, would have been obtained. There are thus three different prism forms, each of which is parallel to one of the crystallographic axes. On a real crystal two or more of these forms could coexist as in Fig. 6.4, but they would still be distinct forms, since the faces in one would not be related by the symmetry to the faces in another.

If the pole is moved to a vertex of the representative triangle, then the face it represents is simultaneously perpendicular to two mirror planes and is not repeated by either of them,

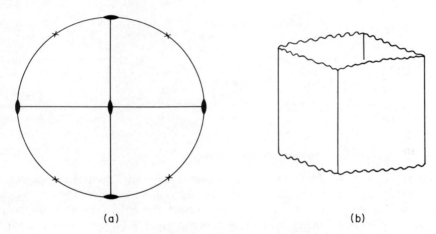

FIG. 6.3. The prism form {hk0} of the orthorhombic system.

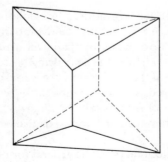

FIG. 6.4. Combination of the {hk0} and {h0l} orthorhombic forms.

but only by the one remaining mirror. This therefore gives a form consisting only of two parallel faces, a pinacoid. There are three such forms, {100}, {010} and {001}, and these again are special forms.

It is not necessary to discuss here in detail the methods of finding the axial ratios of an orthorhombic crystal, as these have been dealt with adequately in preceding chapters. Initially there is freedom of choice in the order in which the three diads are labelled as the x-, y- and z-axes. Then indices are assigned on the basis of this choice, and if we know the orientations of a face in any two of the prism forms {110}, {011} and {101} we can calculate the corresponding two ratios among a/b, b/c and c/a, and hence obtain $a:b:c$. If a face to be used happens to have more general indices like {$hk0$} then it gives initially ha/kb but the principle is unchanged. If the crystal does not possess appropriate prism forms but possesses a general form {hkl} then we can use the orientation of a face in this form to calculate the theoretical position of the corresponding prism faces in the way that was demonstrated on p. 52 for finding (011) from (111).

The morphological forms of the holosymmetric class of the orthorhombic system are summarised below:

Form indices	Shape	No. of faces in form	Nature
{100}, {010}, and {001}	pinacoid	2	open
{$hk0$}, {$0kl$}, and {$h0l$}	prism	4	open
{hkl}	bipyramid	8	closed

The tetragonal system (holosymmetric class)

Here there is a 4-fold rotation axis along z and an "equatorial" mirror plane perpendicular to it. There are also four mirror planes intersecting in the 4-fold axis, and 2-fold rotation axes where each of these intersect the equatorial mirror plane. The relationship of these symmetry elements is shown in Fig. 6.5. The representative triangle is shaded, and its sides are again the representations of mirror planes and its vertices are the representations of rotation axes, but the latter are of two kinds; one is a 4-fold axis and the other two are 2-fold axes.

If the pole of a face is inserted within the representative triangle, then it is repeated sixteen times as shown in Fig. 6.5, eight times on the top of the crystal and eight times on the bottom. The faces on the top occur as four pairs, the members of a pair being related by a mirror plane (e.g. (hkl) and ($h\bar{k}l$)) and the pairs being related to one another by the 4-fold axis. Thus the pair (hkl) and ($h\bar{k}l$) are related by the 4-fold axis to faces with the values of the indices h and k interchanged, which we symbolise by writing (khl) and ($k\bar{h}l$), etc. Such an arrangement of four pairs of faces is called a *ditetragonal pyramid*. The general form then has a ditetragonal pyramid at the top and also at the bottom and is called a ditetragonal bipyramid.

Since the indices of the sixteen faces in this general form {hkl} are derived not only by permuting + and − signs on the indices but also by interchanging the values of the first two indices, it is evident that these two indices cannot be equal if the form is to be a general

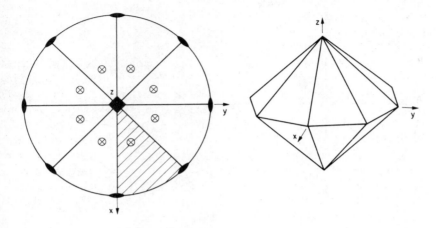

FIG. 6.5. (a) Symmetry and the general form $\{hkl\}$ of the holosymmetric class of the tetragonal system, with one representative triangle shaded. (b) The faces of the form, a ditetragonal bipyramid.

one. A face with the first two indices equal, symbolised as (hhl), would have its pole on the representation of the mirror plane that bisects the angle between the x- and y-axes, and would belong to a special form $\{hhl\}$ containing eight faces, four at the top and four at the bottom, in the shape of a *tetragonal bipyramid* (Fig. 6.6(a)). The indices of the faces in this form are obtained simply by permuting $+$ and $-$ signs. A similar tetragonal bipyramid form, but in a different orientation with respect to the axes, is obtained if we consider the pole of a face lying on the representation of the mirror plane perpendicular to the x- or y-axis. This is the form $\{h0l\}$, whose faces have the indices $(h0l)$, $(0hl)$, $(\bar{h}0l)$, $(0\bar{h}l)$, $(h0\bar{l})$, $(0h\bar{l})$, $(\bar{h}0\bar{l})$ and $(0\bar{h}\bar{l})$ (Fig. 6.6(b)).

If a pole is considered on the third side of the representative triangle its indices will be $(hk0)$, and if $h \neq k$ then this face will be repeated to $(\bar{h}k0)$ by the axial mirror plane, and this pair of faces will be repeated by the 4-fold axis to give the *ditetragonal prism* $\{hk0\}$ with eight faces. If the pole is moved to either of the two vertices of the representative triangle on the primitive then it lies on two mirror planes and corresponds to a *tetragonal prism*, either $\{100\}$ or $\{110\}$, with four faces. Finally, if the pole is moved to the centre of the stereogram (001) it is not repeated by the 4-fold axis but only by the equatorial mirror to $(00\bar{1})$, thereby giving a pinacoid. Such a pinacoid perpendicular to a *principal* axis (i.e. in the tetragonal, hexagonal or trigonal systems) is often called a *basal* pinacoid.

In summary the forms of the tetragonal holosymmetrical class are:

Form indices	Shape	No. of faces in form	Nature
$\{001\}$	pinacoid	2	open
$\{100\}$, $\{110\}$	tetragonal prism	4	open
$\{hk0\}$	ditetragonal prism	8	open
$\{h0l\}$ $\{hhl\}$	tetragonal bipyramid	8	closed
$\{hkl\}$	ditetragonal bypyramid	16	closed

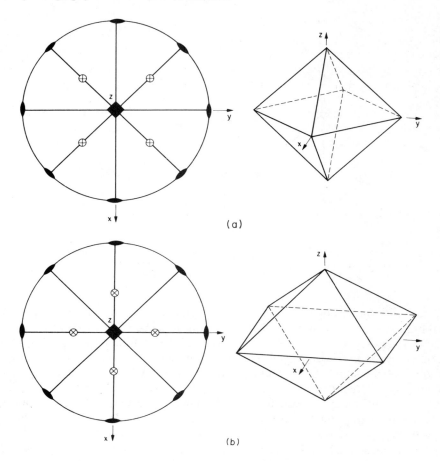

FIG. 6.6. (a) The form $\{hhl\}$ and (b) the form $\{h0l\}$ of the holosymmetric class of the tetragonal system, tetragonal bipyramids in different orientations.

In indexing such a tetragonal crystal the z-axis is uniquely defined by the direction of the tetrad, and since the x- and y-axes are related by the symmetry no difference in nomenclature arises from the order in which they are chosen. However, because there are two pairs of orthogonal diads perpendicular to z- there is an arbitrary choice as to which of these pairs shall be taken as the axial directions. This choice defines whether a particular tetragonal bipyramid form will be indexed as $\{101\}$ or as $\{111\}$, and leads to values of the ratio c/a that differ by a factor of $\sqrt{2}$.

The cubic system (holosymmetric class)

In this system the crystallographic axes are defined unambiguously by the symmetry, and they coincide with the three 4-fold axes that are present in the holosymmetric class.

Since the edges of the unit cell are all equal there is no axial ratio to be determined, and the orientation of a face with given indices is determinate, and the same for every cubic crystal. There are three axial mirror planes and six diagonal ones, and these divide up the sphere into forty-eight representative triangles, the twenty-four that lie on the top of the sphere being shown in the stereographic representation of the symmetry in Fig. 6.7.

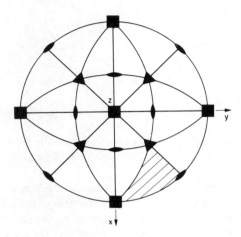

FIG. 6.7. Stereographic representation of the symmetry of the cube, with a representative triangle shaded.

Although the triads in the symmetry of the cube are strictly $\bar{3}$-axes, it simplifies the consideration of the symmetry to regard them as ordinary 3-fold rotation axes, and they are shown as such in Fig. 6.7. This is permissible since a $\bar{3}$-axis is equivalent in its effects to a 3-fold rotation axis together with an independent centre of inversion ($\bar{1}$), and the latter is necessarily implied by the intersection of the axial mirror planes. It is to be noted that the 3-fold axes lie at the centres of the octants between the axes and relate the cubic axes and the axial mirror planes to one another (respectively).

The vertices of a representative triangle (shown shaded) lie on a 4-fold axis, a 3-fold axis and a 2-fold axis, and the three sides of the triangle are therefore different from one another, lying as they do between different pairs of these three different kinds of symmetry elements. There are therefore seven essentially different placings of a pole with respect to the representative triangle – at each vertex, on each side, and within the triangle – and accordingly there are seven different kinds of form. Because of the large number of faces in the general form and in some of the special forms their shape will be more readily appreciated if we proceed from the simplest, most special forms towards the general one.

If a pole is placed on a 4-fold axis, say at the positive end of the x-axis thereby giving it the indices (100), then it is repeated by a 3-fold axis to (010) and (001) and then these are repeated by the axial mirrors (or by other equivalent operations) to ($\bar{1}$00), (0$\bar{1}$0) and 00$\bar{1}$), that is to the ends of all the 4-fold axes in Fig. 6.7. No further repetition occurs, so we have the cube form {100} (Fig. 6.8). Note that in this system, because of the equivalence of all three axes, we always have complete permutation of the three indices (in so far as they are

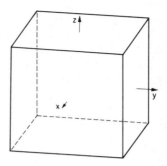

FIG. 6.8. The cube form {100}, with the directions of the axes indicated.

numerically different) in generating the face indices from the form indices. The {100} form is frequently exhibited as the dominant form, or alone, by many cubic crystals, e.g. fluorite, galena, halite.

If a pole is placed on a 3-fold axis in Fig. 6.7, then the corresponding face must be equally inclined to the three axes, and in view of the equal lengths of the edges of the cubic unit cell it must have the indices (111). It is evidently repeated by the symmetry to both ends of all the 3-fold axes since these are symmetrically equivalent; this can be done in many ways, perhaps most simply by considering the effect of the axial mirror planes. The eight faces in the form {111} therefore have indices derived from all permutations of + and − signs applied to the three unit indices. The shape of the form is the regular octahedron, and its relationship to the cube is shown in Fig. 6.9 by a crystal exhibiting the cube and octahedron forms together. The octahedron truncates the vertices of the cube. The form {111} again is frequently exhibited as the dominant form, or alone, by many cubic crystals, notably by the various members of the spinel group.

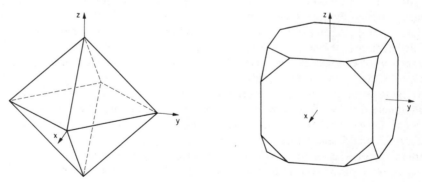

FIG. 6.9. The octahedron {111}, alone and in combination with the cube {100}.

If the pole is placed on a 2-fold axis in Fig. 6.7 then clearly it must be repeated at both ends of all six 2-fold axes, since they are symmetrically equivalent. This repetition can probably be envisaged most simply in terms of the operation of one of the 3-fold axes followed by the axial mirror planes. Since the 2-fold axes bisect the angles between the crystallographic axes taken in pairs, the indices of the face of this form whose pole lies between the x- and y-axes on the primitive is (110). The form is accordingly {110}, and the indices of the other faces are obtained by permuting the position of the zero and of + and − signs on the two non-zero indices. The form is the rhombic dodecahedron having twelve rhombus faces. Its relationship to the cube and octahedron is shown in Fig. 6.10; it truncates the edges of both of them and provides a good example of the variability of the shape of faces of a given form depending on the form with which it is combined. The {110} form is occasionally exhibited as a dominant form, or alone, especially by garnets; it is more usual as a minor form in association with the cube, or with the octahedron as in magnetite.

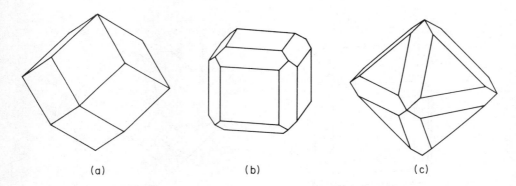

FIG. 6.10. The rhombic dodecahedron {110}: (a) alone, (b) in combination with the cube {100}, and (c) in combination with the octahedron {111}.

Having considered the effect of putting a pole at each of the vertices of the representative triangle, we now consider putting one on a side, starting with the side joining a 4-fold axis to a 2-fold axis. Such a face with its pole on the primitive and near the positive x-axis has the indices $(hk0)$ with $h > k$. If this is operated on by the 4-fold axis we obtain a set of four faces around the positive x-axis, and this set will then clearly repeat around the +ve and −ve ends of all the 4-folds axes (Fig. 6.11), giving twenty-four faces in all. These will include all permutations of the order of h, k and 0, as well as th signs of h and k; thus although we started with $(hk0)$ with $h > k$ the form will include $(kh0)$, that is a face having its first index greater than its second one. The shape is like that of a cube with a low square pyramid erected on each face, the inclination of these faces depending of course on the ratio of h/k (Fig. 6.11 is drawn for the form {310}). The faces of this form bevel the edges of the cube; in combination with the octahedron they modify the vertices with sets of four faces forming a less acute point than that of the octahedron itself and twisted through 45° with respect to it. The $\{hk0\}$ form very rarely occurs alone in holosymmetric cubic crystals, but it occurs in combination with the cube, notably in fluorite.

A pole placed between the 3-fold axis (at the position of (111)) and the 2-fold axis (at the position of (110)) will have indices which are linear combination of (110) and (111) (see p. 34); that is its first and second indices will be equal, and since it is further from the z-axis

62 Crystallography for Earth Science Students

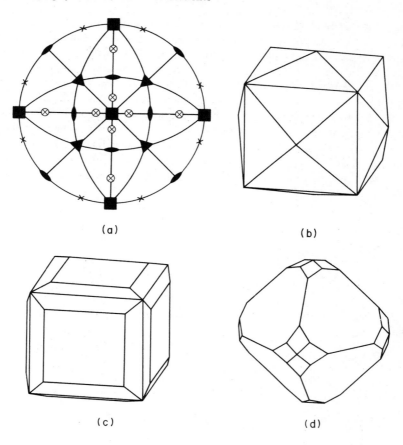

FIG. 6.11. The {310} form of the holosymmetric class of the cubic system. (c) and (d) show combinations with {100} and {111}.

than from the x- and y-axes its third index will be smaller than these. We symbolise this situation as (hhl) with $h > l$, meaning that the repeated index is greater than the other, and this remains true for all the faces of the form; the alternative with the repeated index less than the other would place the pole on the same great circle but on the far side of the 3-fold axis as seen from the 2-fold axis. It would therefore be between a 3-fold axis and a 4-fold axis and would not be a member of the form under consideration. The pole (hhl) is repeated around the 3-fold axis (Fig. 6.12), giving rise to (lhh) and (hlh), and this set of three faces is then repeated into all the other octants by the axial mirrors, involving permutations of $+$ and $-$ signs for the three indices. Thus there are again twenty-four faces in the form, which has the shape of an octahedron with a low trigonal pyramid erected on each face. The faces of this form bevel the edges of the octahedron, and modify the vertices of the cube into less acute points, as shown by the drawing of the particular case {221} in Fig. 6.12.

Morphology of the Seven Crystal Systems   63

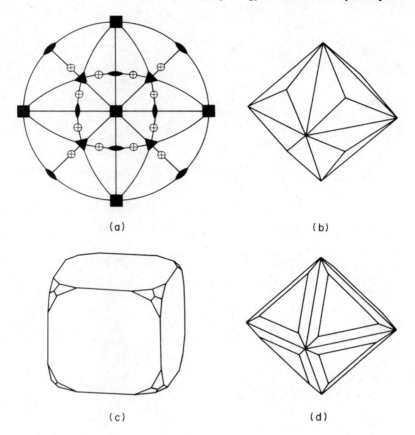

FIG. 6.12. The {221} form of the holosymmetric class of the cubic system. (c) and (d) show combinations with {100} and {111}.

The third edge of the representative triangle is that between a 4-fold axis and a 3-fold axis. Taking such an edge between the positions of (100) and (111) leads to the indices (*hll*) with $h > l$ on the same principles as were used above (Fig. 6.13(a)). This face is again repeated around the 3-fold axis to (*lhl*) and (*llh*), and then by the axial mirrors into the other quadrants involving permutation of + and − signs on the three indices. Again there are therefore twenty-four faces in the form, but it has "kite-shaped" instead of triangular faces, and it cannot be described so simply in relation to the cube or octahedron since it does not have a set of edges that corresponds to the edges of either of these figures, as may be seen from Fig. 6.13(b). Its faces modify the vertices of both the cube and the octahedron, as shown for the particular case of the form {211}.

If a pole is placed in a totally general position within a representative triangle (say between (100), (110) and (111)), then it is repeated by a diagonal mirror plane into an adjacent triangle (e.g. that between (100), (101) and (111)); and the resulting pair of faces is

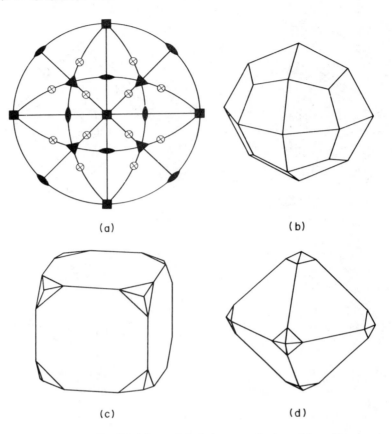

FIG. 6.13. The {211} form of the holosymmetric class of the cubic system. (c) and (d) show combinations with {100} and {111}.

then repeated around the 3-fold axis to give a set of six faces which are further repeated by the axial mirror planes into all the octants. The form {*hkl*} therefore has forty-eight faces (Fig. 6.14), and is most easily understood in terms of a rhombic dodecahedron with a low (rhombic) pyramid erected on each face, as shown for the particular form {321} in Fig. 6.14.

The forms of the holosymmetric class of the cubic system may now be summarised. They are all closed forms.

Form indices	Shape	No. of faces
{100}	cube	6
{111}	octahedron	8
{110}	rhombic dodecahedron	12
{*hk*0}	tetrahexahedron	24
{*hhl*} $h > l$	trisoctahedron	24
{*hll*} $h > l$	24-hedron	24
{*hkl*}	48-hedron	48

Morphology of the Seven Crystal Systems 65

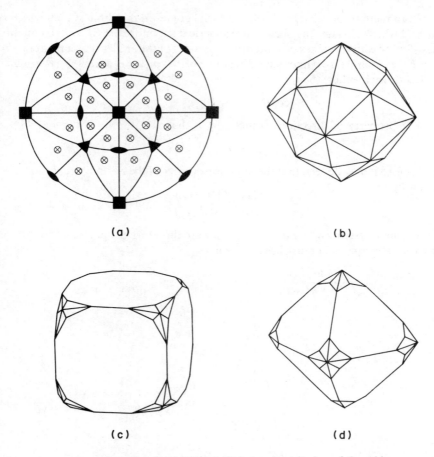

FIG. 6.14. A general form {321} of the holosymmetric class of the cubic system. (c) and (d) show combinations with {100} and {111}.

There are of course an indefinite number of forms with specific values of h, k and l where these are not fixed. The traditional names of $\{hk0\}$ and $\{hhl\}$ are included above for reference because they are descriptive of their relationship to the cube (hexahedron) and octahedron, respectively, but those of $\{hll\}$ and $\{hkl\}$ tend if anything to mislead, and are replaced by the simple terminology of n-hedron.

Now that we have worked up from the most special to the most general form it is instructive to look at the relationships of the forms the other way round. The general form has forty-eight faces. If a pole is moved on to an edge of the representative triangle the corresponding face is perpendicular to a mirror plane and is not repeated by it, so the number of faces is divided by two to give twenty-four. If it is now put perpendicular to a rotation axis it is not repeated by that axis, and the number of faces is divided again by the order of the axis to give twelve (for $\{110\}$), eight (for $\{111\}$) or six (for $\{100\}$) as the case may be.

There is a method of calculation of interfacial angles in the cubic system which is very simple and which obviates the need to use spherical trigonometry. It is based on direction cosines. The direction cosines of a line are the cosines of the angles that it makes with three orthogonal axes, and if two lines have direction cosines p_1, q_1, r_1 and p_2, q_2, r_2 and make an angle θ with each other, then

$$\cos \theta = p_1 p_2 + q_1 q_2 + r_1 r_2. \tag{6.1}$$

It is also a property of direction cosines p, q, r, that

$$p^2 + q^2 + r^2 = 1. \tag{6.2}$$

From Fig. 6.15 it may be seen that the direction cosines of the normal to a crystal face (hkl) are given by

$$\frac{ON}{a/h}, \frac{ON}{b/k}, \frac{ON}{c/l}$$

and in the cubic system (where $a = b = c$) they are therefore proportional to h, k, l and in order to satisfy eqn. (6.2) must have the values

$$\frac{h}{\sqrt{(h^2 + k^2 + l^2)}}, \frac{k}{\sqrt{(h^2 + k^2 + l^2)}}, \frac{l}{\sqrt{(h^2 + k^2 + l^2)}}.$$

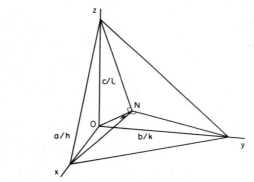

FIG. 6.15. Direction of the normal ON to a face (hkl) in an orthogonal system.

Thus the angle between any two faces $(h_1 k_1 l_1)$ and $(h_2 k_2 l_2)$ can be very easily calculated from eqn. (6.1) in the form

$$\cos \theta = \frac{h_1 h_2 + k_1 k_2 + l_1 l_2}{\sqrt{(h_1^2 + k_1^2 + l_1^2)} \sqrt{(h_2^2 + k_2^2 + l_2^2)}}.$$

The monoclinic system (holosymmetric class)

By contrast with the cubic system where the complications are due to an excess symmetry, in the monoclinic system there is very little symmetry, but quite different problems arise from the non-orthogonality of the axes.

In representing monoclinic crystals by means of a stereogram it is usual to retain the convention adopted in other systems of putting the z-axis vertical so that it emerges in the centre of the stereogram. The 2-fold axis (conventionally along the y-axis) therefore runs from right to left as in Fig. 6.16(a) and the mirror plane is represented by the diameter perpendicular to this. Since the x-axis makes an obtuse angle β with the z-axis it does not lie in the plane of the primitive but emerges in the southern hemisphere, its projection on that plane lying along the broken line in Fig. 6.16(a). Although this is in general the preferred method of representation of the monoclinic system it is sometimes more convenient to adopt that shown in Fig. 6.16(b), with the y-axis (and the diad) vertical and emerging in the centre of the projection. The mirror plane is then the plane of the primitive, which also contains the x- and z-axes.

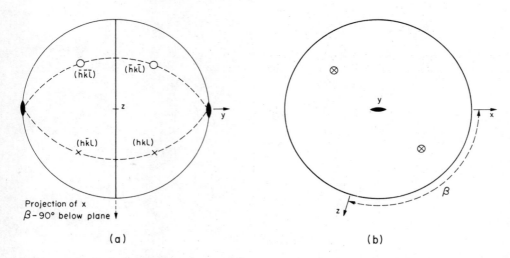

FIG. 6.16. Stereogram of the symmetry and the general form $\{hkl\}$ of the holosymmetric class of the monoclinic system: (a) conventional orientation, (b) alternative orientation.

From Fig. 6.16 it can be seen that there is no such thing as a representative triangle. A general face (hkl) is repeated by the 2-fold axis to $(\bar{h}k\bar{l})$, and these faces are repeated by the mirror to $(h\bar{k}l)$ and $(\bar{h}\bar{k}\bar{l})$. The general form $\{hkl\}$ is therefore an open prism form containing these four faces, and with its prism axis perpendicular to the y-axis but otherwise undefined. This is most readily appreciated from Fig. 6.16(b). If the pole is placed either on the mirror plane to give $\{h0l\}$, or on the 2-fold axis to give $\{010\}$, the number of faces is reduced by a factor of two to a pinacoid. Thus the only forms are as follows, and they are all open:

Form indices	Shape	No. of faces
$\{010\}$	pinacoid	2
$\{h0l\}$	pinacoid	2
$\{hkl\}$	prism	4

It is to be noted that (100) and (001) do not in this system constitute separate kinds of form; they are just particular cases of $\{h0l\}$. Similarly $\{hk0\}$ and $\{0kl\}$ are just particular cases of the general form. Also, the position of the faces (100) and (001) on the stereogram are rather different from what we are used to, because they do not coincide with the points of emergence of the x- and z-axes. In the standard orientation of the stereogram the pole of (100) lies on the primitive, since this face is parallel to the y- and z-axes, and the pole of (001) lies at an angle β-90° from the centre in order to be 90° from the point of emergence of the x-axis (Fig. 6.17). Clearly the angle between (100) and (001) is $180° - \beta$, and this usually provides the most convenient way of finding β. If these two faces are not themselves present their theoritical positions can always be derived as the positions where appropriate zone circles cross the mirror plane.

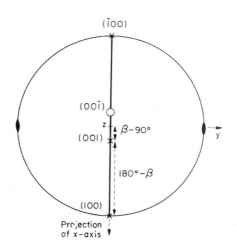

FIG. 6.17. Positions of (100), ($\bar{1}$00), (001) and (00$\bar{1}$) on the monoclinic stereogram.

Once β is known we can find $a:b$ from the position of (110). If, for example, the angle (θ) between (110) and (1$\bar{1}$0) is known, we may proceed by the method shown in Fig. 6.18(a). In the triangle OAB the angle B is $\theta/2$. Thus

$$OA/OB = \tan \tfrac{1}{2}\theta.$$

However, there is a difference from corresponding calculations in the orthorhombic system in that, whereas OB is the intercept b of (110) on the y-axis, OA is not the intercept on the x-axis but on the projection of that axis on the plane of the paper. The axis itself is at an angle β-90° below the plane of the paper and so

$$OA = a \cos(\beta - 90°).$$

Therefore
$$a:b = \tan \tfrac{1}{2}\theta : \cos(\beta - 90°).$$

Exactly similar considerations would apply in deriving $b:c$ from the position of (011). On the other hand, if we wish to derive $a:c$ from the position of (101) it is more convenient to

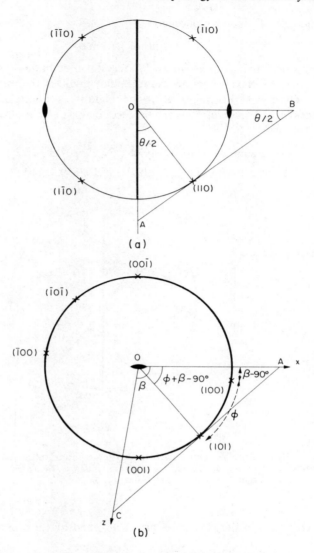

FIG. 6.18. Triangles on the stereogram to calculate the axial ratios of a monoclinic crystal.

consider a stereogram in the alternative orientation shown in Fig. 6.18(b), with the diad in the centre. In this stereogram both the x- and z-axes lie in the plane of the paper. If (101) is known to make an angle ϕ with (100) then the angles of the triangle OAC are

$$O = \beta,$$
$$A = 90° - (\phi + \beta - 90°)$$
$$= 180° - \phi - \beta,$$

and hence $$C = \phi.$$

70   Crystallography for Earth Science Students

Thus by the sine rule of plane trigonometry

$$a:c = \sin(\phi + \beta):\sin\phi.$$

The general approach to indexing a monoclinic crystal and obtaining its axial ratios and interaxial angle may be clarified by an example that involves the minimum necessary number of faces. The crystal shown in Fig. 6.19 has eight faces that clearly lie in two prism zones of four faces each. The angles between the faces in these two zones are measured and found to be as follows:

in the zone MNPQ   107.8°, 72.2°, 107.8°, 72.2°;
in the zone RSTU   139.9°, 40.1°, 139.9°, 40.1°.

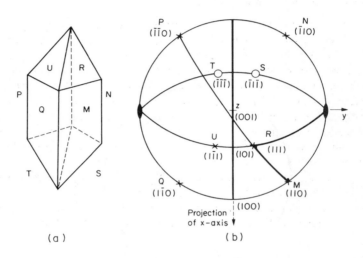

FIG. 6.19. (a) A monoclinic crystal showing two prism forms. (b) Stereogram of (a) showing the indexing of the faces and constructions to derive by spherical trigonometry the angles required to find $a:b:c$.

The relative orientations of the two zones are established by the angles

$$MR = NS = 52.3° \quad \text{and} \quad QR = PS = 78.0°.$$

The two prisms are evidently general forms, but we may arbitrarily take MNPQ to be the form $\{110\}$, thereby defining the direction of the z-axis, and enabling us to plot this zone round the primitive. The values of MR and QR locate R. Since T is evidently opposite to R it can be located, and NS and TS then locate S, and U is opposite to it. From the stereogram thus constructed (Fig. 6.19(b)) the direction of the y-axis is defined since R and S are clearly related by a 2-fold rotation axis, and R and U by a mirror. The form RSTU can be arbitrarily designated $\{111\}$, another general form in the monoclinic system. The theoretical position of (001) can now be found by drawing the great circle through (110) and (111), since it lies on the mirror plane and in the zone containing these two faces. The theoretical position of (101) can be found similarly from the zone through (111) and ($1\bar{1}1$).

Morphology of the Seven Crystal Systems   71

The whole problem can now be solved by spherical trigonometry. The sides of the spherical triangle YMR are known (MY = ½MN, RY = ½RS) so the angle at Y can be found from the cosine rule, and the angle at M can be found from the sine rule. In the right-angled triangle with its vertices at (001), (100) and (110), the side (100)–(110) is ½QM and the angle at M is the supplement of that found from the triangle YMR; it is therefore possible to use Napier's rule to find the side (001)–(100), which is $180°-\beta$ and gives the value of β. The angle at Y found from the triangle YMR is identified with (100)^(101), so that the ratio of $a:c$ can be calculated in the way already discussed, as also can $a:b$ using the value of YM and the value of β. Thus all the parameters $a:b:c$ and β are obtainable.

The triclinic system (holosymmetric class)

In the triclinic system there is so little symmetry that there is virtually nothing to say about forms. The holosymmetric class has a centre of inversion ($\bar{1}$), so every face (hkl) is repeated to an opposite face ($\bar{h}\bar{k}\bar{l}$) to give a pinacoid, and there are no special forms. The minimum number of forms that can bound a crystal is three, and the shape of such a crystal is entirely dependent on the shape of the unit cell. Moreover, if only three forms were present these could be arbitrarily labelled {100}, {010} and {001} which would be sufficient to define the interaxial angles but would provide no information about the axial ratios; for these to be obtainable at least one more form would have to be present whose faces did not lie in a zone between any pair of the others. Such a form could then be labelled {111} and would permit the axial ratios to be determined.

In constructing a triclinic stereogram (Fig. 6.20) there is no advantage to be gained from departing from the convention that the z-axis emerges in the centre of the stereogram. The angle β is conventionally made obtuse, so the x-axis is in exactly the same position below the plane of the paper as in the monoclinic system. There is no very firm convention as to

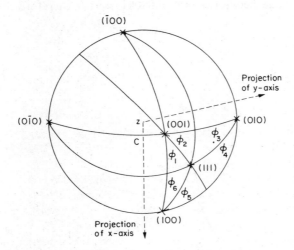

FIG. 6.20. Stereogram of a triclinic crystal showing the spherical triangles required for calculation of $a:b:c$ and α, β, γ.

whether α is obtuse, so the y-axis may be either above or below the plane of the paper, and of course it will also be inclined to the horizontal line on the page. None of the faces (100), (010) or (001) will coincide with the point of emergence of an axis, but (100) and (010) still lie on the primitive, and they will be 90° from the projections of the y- and x-axes respectively.

In order to determine the unit cell it suffices to determine five angles. The angle (100) – (010) permits these two faces to be placed appropriately on the primitive. The angles (100) – (001) and (010) – (001) then serve to locate (001), and two further angles between (111) and any two of (100), (010) and (001) suffice to locate (111). If great circles are drawn through all six pairs of faces chosen from these four (Fig. 6.20) it is then possible to solve appropriate spherical triangles to find the three interaxial angles and the axial ratios. The great circle through (010) and (001) intersects that through x and z at the point C. Since (001) is parallel to the x-axis the latter must be 90° from C and so

$$\beta = 90° + Cz$$
$$= 90° + 90° - (\phi_3 + \phi_4)$$
$$= 180° - (\phi_3 + \phi_4)$$

using the nomenclature of Fig. 6.20.

Similarly $\quad\quad \alpha = 180° - (\phi_5 + \phi_6)$

and $\quad\quad \gamma = 180° - (\phi_1 + \phi_2)$.

But $(\phi_1 + \phi_2)$, etc., are the angles of the spherical triangle (100), (010), (001) whose sides are known, and therefore α, β, γ may be obtained by solving this triangle. In order to see how to find the axial ratios it is useful to draw a diagram of the three axes intersected by the face (111) as in Fig. 6.21. Then it may be seen that in the plane triangle OLM the sine rule gives.

$$a : b = \sin O\hat{M}L : \sin O\hat{L}M.$$

Also OM is the edge between (100) and (001) and is therefore the axis of the zone

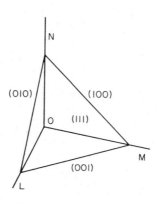

FIG. 6.21. The face (111) relative to the oblique axes of a triclinic crystal. The axial planes are parallel to (100), (010) and (001), and are labelled with these indices.

containing them; similarly OL is the axis of the zone containing (010) and (001), and LM is the axis of the zone containing (111) and (001). Since the angle between the two zone axes is the same as the angle between the great circles representing the two zones, we may then identify angle OML of Fig. 6.21 with ϕ_1 of Fig. 6.20, and by similar reasoning angle OLM can be identified with ϕ_2. Thus

$$a:b = \sin\phi_1 : \sin\phi_2.$$

Similarly
$$b:c = \sin\phi_5 : \sin\phi_6.$$

If (111) was positioned by the angles (100)^(111) and (001)^(111) the required values of ϕ_1 and ϕ_6 may be obtained by solving the spherical triangle (100), (001), (111). Then since $(\phi_1 + \phi_2)$ and $(\phi_5 + \phi_6)$ are already known the problem is solved. If a different pair of angles was measured in order to locate (111) it will always be possible to use one of these two relationships together with the third analogous one

$$c:a = \sin\phi_3 : \sin\phi_4.$$

This method has been developed here for the triclinic system where no short cuts are possible. It is of course completely general and can be applied equally to any of the other systems. It will in fact reduce to the special methods we have developed previously when account is taken of the 90° angles that are present in the more symmetrical systems.

The hexagonal system (holosymmetric class)

In Chapter 3 it was demonstrated that although a unit cell in the form of a parallelepiped cannot have a 6-fold rotation axis a stack of such cells can constitute a lattice with this symmetry (Fig. 3.5(g)). They then have a and b equal to each other and to the short diagonal of the ab face, $\alpha = \beta = 90°$, and $\gamma = 120°$. Such a hexagonal lattice has the symmetry shown in Fig. 6.22, where the 6-fold axis (the z-axis) emerges at the centre of the stereogram in the usual way.

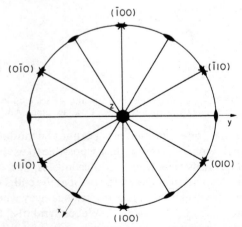

FIG. 6.22. Stereogram of the full symmetry of the hexagonal system, and a face (100) repeated round the primitive by the 6-fold axis and indexed with ordinary Miller indices.

In this system a slight modification of the usual system of Miller indices is found to be desirable. The reason for this can be seen from Fig. 6.22, where a hexagonal prism form has been generated by inserting the (100) face 90° from the y-axis, on the primitive, and subjecting it to the symmetry operation of the 6-fold axis. The result is obviously a hexagonal prism form which we would expect to call {100}, but the indices of the faces are (100), (010), ($\bar{1}$10), ($\bar{1}$00), (0$\bar{1}$0) and (1$\bar{1}$0). Thus they cannot all be derived by permuting the values and signs of the indices of the first-named face. By comparison with the situations in the systems considered previously it is anomalous that ($\bar{1}$10) and (1$\bar{1}$0) should belong to the form {100}, and there is no obvious numerical rule for generating the indices of faces in the form. The same problem arises in all other forms of the system except for {001} where repetition by the 6-fold axis is not involved. It is overcome by the introduction of a fourth axis (w) in the plane of the x- and y-axes. The positive ends of the x-, y- and w-axes thus make angles of 120° with each other (Fig. 6.23). The w-axis is redundant, in the sense that positions along it can be specified in terms of x- and y-coordinates. Its value is that any plane which has intercepts a/h on the x-axis and a/k on the y-axis also has an intercept $-a/(h+k)$ on the w-axis, so that we can define a fourth index $i = -h - k$, and the indices ($hkil$) of a plane (its Miller–Bravais indices) then permute in the usual way as a result of the operation of the 6-fold symmetry. For example, the form discussed above and depicted again in Fig. 6.24 is the {10$\bar{1}$0} form, and its component faces are (10$\bar{1}$0), (01$\bar{1}$0), ($\bar{1}$100), ($\bar{1}$010), (0$\bar{1}$10) and (1$\bar{1}$00), where the positions and signs of the first three indices permute in all possible ways that are consistent with the relation $i = -h - k$.

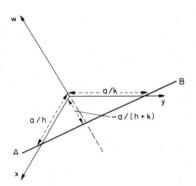

Fig. 6.23. The three coplanar x-, y- and w-axes used in the hexagonal system, intercepted by a face along AB.

It is important to note that zone symbols [uvw] cannot be modified in the same simple way to deal with hexagonal symmetry. It is perfectly possible to use zone symbols [uvw] in the ordinary way. By temporarily reverting to Miller indices (hkl) one can simply ignore the i-index during a zonal calculation and then restore it at the end. However, if it is desired to have zone symbols that themselves permute in the usual way when operated on by the hexagonal symmetry, then use may be made of Weber symbols [$UVJW$] defined as

$$U = u - (u+w)/3, \quad V = v - (u+v)/3, \quad J = -(U+V),$$
$$W = w \text{ if } u+v = 3n, \quad \text{or} \quad W = 3w \text{ if } u+v \neq 3n.$$

From Fig. 6.24 it can be seen that the representative triangle in the hexagonal stereogram is bounded by the equatorial mirror plane on the primitive and two mirror planes intersecting at 30° in the z-axis; its vertices are at the positions of the 6-fold axis and two 2-fold axes.

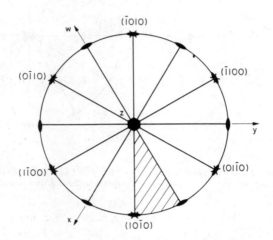

FIG. 6.24. As Fig. 6.22 but with Miller–Bravais indices. A representative triangle is shaded.

If we place a pole (0001) at the position of the 6-fold axis it is not repeated by it, but only by the equatorial mirror to (000$\bar{1}$). Thus the form {0001} is the basal pinacoid. A pole placed at the vertex between the x-axis and the $-w$-axis gives the result shown in Fig. 6.24, which we have already seen leads to the hexagonal prism form {10$\bar{1}$0}. Choice of the vertex on the w-axis also obviously gives a hexagonal prism, but its indices are {11$\bar{2}$0}, since it includes a face making equal angles with the x- and y-axes.

A face in a general position on the primitive (hki0) is repeated by the mirror that bisects the angle between x and y to (khi0), and this pair of faces is then repeated by the 6-fold axis to give a dihexagonal prism (Fig. 6.25).

A pole between (10$\bar{1}$0) and (0001) is ($h0\bar{h}l$), and the form $\{h0\bar{h}l\}$ is obviously a hexagonal bipyramid, and a second such form $\{hh\overline{2h}l\}$ is obtained by starting with a pole between (1120) and (0001) (Fig. 6.26). A pole in a completely general position ($hkil$) lies between (hki0) and (0001) and clearly gives a dihexagonal bipyramid (Fig. 6.27). In summary the forms of the holosymmetric class of the hexagonal system are:

Form indices	Shape	No. of faces in form	Nature
{0001}	pinacoid	2	open
{10$\bar{1}$0}, {11$\bar{2}$0}	hexagonal prism	6	open
{hki0}	dihexagonal prism	12	open
$\{h0\bar{h}l\}$ $\{hh\overline{2h}l\}$	hexagonal bipyramid	12	closed
{$hkil$}	dihexagonal bipyramid	24	closed

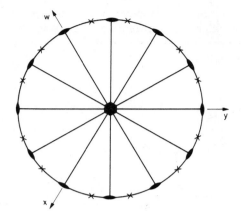

FIG. 6.25. The form $\{hki0\}$, a dihexagonal prism in the holosymmetric class of the hexagonal system.

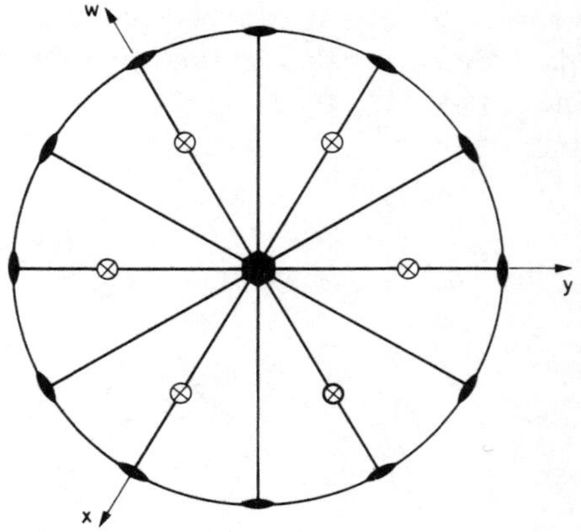

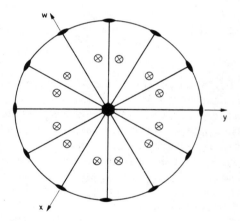

FIG. 6.26. The form $\{hh\overline{2h}l\}$, a hexagonal bipyramid in the holosymmetric class of the hexagonal system.

FIG. 6.27. The general form $\{hkil\}$, a dihexagonal bipyramid in the holosymmetric class of the hexagonal system.

Thus there is a very close analogy between the forms of the tetragonal and hexagonal systems, the only difference being that due to the change from 4-fold to 6-fold symmetry.

In the hexagonal system there is only one unit cell parameter to determine, namely c/a. This can most simply be calculated from the angle θ between $(10\overline{1}0)$ and $(10\overline{1}1)$, since the face $(10\overline{1}1)$ has an intercept (ON in Fig. 6.28) of $a \cos 30°$ on the normal to $(10\overline{1}0)$ and c on the z-axis. Thus

$$c/a = \cos 30°/\tan \theta.$$

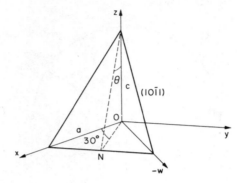

Fig. 6.28. Intersection of the x- and z-axes by the face (10$\bar{1}$1).

Alternatively the face (11$\bar{2}$1) has an intercept of c on the z-axis and $a/2$ on the $-w$-axis, so that if the angle θ is between (11$\bar{2}$1) and (11$\bar{2}$0) then

$$c/a = \tfrac{1}{2}\tan\theta.$$

If a general form is present, but neither of the bipyramid forms, then the positions of the latter can be derived by drawing in appropriate zones on the stereogram.

The trigonal system (holosymmetric class)

In Chapter 3 the unit cell of the trigonal system was defined as a rhombohedron having a triad (actually a $\bar{3}$-axis of rotation–inversion) along one of its diagonals, so that its three edges are equal and its three interaxial angles are also equal. It has three mirror planes intersecting along the 3-fold axis and each containing two polar edges of the rhombohedron, and three 2-fold rotation axes joining mid-points of opposite equatorial edges. The natural orientation of a stereogram to display this symmetry is that shown in Fig. 6.29. The x, y and z-axes of the cell then emerge at three symmetrically related points on the stereogram that all correspond to points on the upper half of the sphere.

This description of the axes of the trigonal system turns out to be very undesirable when it comes to index the forms. If we put a pole at the intersection of a mirror plane with the primitive it is repeated by 2-fold axes to positions every 60° round the primitive, thereby giving a hexagonal prism. This is a special form in the trigonal system, but the indices of the six faces involved can be shown to be (2$\bar{1}\bar{1}$), (11$\bar{2}$), ($\bar{1}$2$\bar{1}$), ($\bar{2}$11), ($\bar{1}\bar{1}$2) and (1$\bar{2}$1) which do not look to be of a very special type; they are certainly very different from those of an axial prism in any other system, and give rather little direct evidence of the character of the form. Some other special forms (such as $\{41\bar{2}\}$) have even stranger looking indices. It is therefore usually more convenient to adopt a different notation.

It may be seen by a comparison of Fig. 6.29 with Fig. 6.22 that there is a close relationship between the symmetry of the trigonal and hexagonal systems. The former is obtained from the latter by reducing the 6-fold axis to 3-fold, and by omitting the

78   Crystallography for Earth Science Students

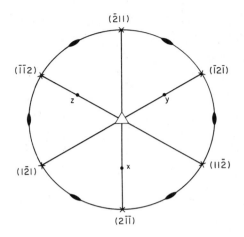

Fig. 6.29. Symmetry of the holosymmetric class of the trigonal system showing the orientations of the x-, y- and z-axes of the rhombohedral cell. A prism form is shown with indices referred to the rhombohedral axes.

equatorial mirror plane and alternate axial mirror planes. The omission of the equatorial mirror necessarily implies the loss of the 2-fold axes where the remaining axial mirror planes intersect the primitive. This reduction of symmetry is very similar to that which occurs in passing from the holosymmetric class of a system to a class of lower symmetry (which will be discussed in Chapter 7), and it is in fact possible to treat the trigonal system on the basis of a lattice having a unit cell of the same shape as that of the hexagonal system, but the contents of which are of lower symmetry. Some reference was made in Chapter 3 to the relationship between the trigonal and hexagonal systems, and it will be clarified further in Chapter 10; for the present purpose it suffices to observe that it is always possible to choose a cell of shape appropriate to the hexagonal system in a lattice of trigonal symmetry, and we shall discuss the morphology of the trigonal system in terms of hexagonal indexing.

To do this it is most convenient to take the pole of a face in each of the forms in the hexagonal system and consider how it is repeated by the reduced symmetry of the trigonal system. Thus we can find out whether the form is the same, and, if not, how it is modified. Those forms that are not modified are shown in Fig. 6.30(a) and (b).

$\{0001\}$ Although (0001) is no longer repeated to $(000\bar{1})$ by an equatorial mirror, it is still related to that face by the 2-fold axes, or by the inversion inherent in the $\bar{3}$-axis.

$\{10\bar{1}0\}$ Although there is no 6-fold axis, $(10\bar{1}0)$ is still repeated into a hexagonal prism by the 2-fold axes at every 60° round the primitive.

$\{11\bar{2}0\}$ This is also still repeated into a hexagonal prism, but this time by the mirror planes disposed every 60° around the z-axis.

$\{hki0\}$ The face $(hki0)$ is repeated by a 2-fold axis to $(khi0)$ and this pair of faces then undergoes 6-fold repetition by the mirrors, so that it still gives a dihexagonal prism.

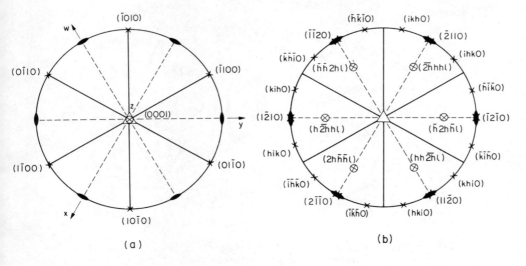

FIG. 6.30. Forms of the holosymmetric class of the trigonal system (indexed on hexagonal axes) that are identical with the corresponding hexagonal forms. Faces with l negative are plotted on the stereogram but not labelled separately from the corresponding face with l positive.

$\{hh\bar{2}hl\}$ The face $(hh\bar{2}hl)$ is repeated by a 2-fold axis to $(hh\bar{2}h\bar{l})$, and this pair is repeated by the mirror planes to give a hexagonal bipyramid.

All the above five forms are therefore the same as in the hexagonal system. The remaining two forms are modified by the reduction in the symmetry and are shown in Figs. 6.31 and 6.32.

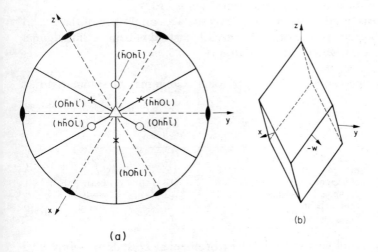

FIG. 6.31. The form $\{h0\bar{h}l\}$ in the holosymmetric class of the trigonal system is a rhombohedron.

80 Crystallography for Earth Science Students

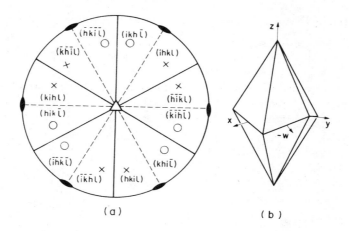

FIG. 6.32. The general form {hkil} in the holosymmetric class of the trigonal system is a ditrigonal scalenohedron.

{hOh̄l} The face (hOh̄l) is repeated by a 2-fold axis to (Ohh̄l̄), but is not repeated to (Ohh̄l). Thus there are only three faces in this form on the top of the crystal and three more on the bottom; instead of being a hexagonal bipyramid it is a rhombohedron (Fig. 6.31(b)) containing the faces (hOh̄l), (Ohh̄l̄) (h̄hOl), (h̄Oh̄l̄), (Oh̄hl̄) and (hh̄Ol).
The other six faces that were present in the hexagonal bipyramid constitute another rhombohedral form {Ohh̄l}. It is to be noted that the form having its faces parallel to those of the rhombohedral cell is {101̄1}.

{hkil} For the same reason this form contains only six faces on the top of the crystal and six on the bottom instead of twelve on each; instead of being a dihexagonal bipyramid it is a ditrigonal scalenohedron (Fig. 6.32).

The forms of the holosymmetric class of the trigonal system may be summarised as:

Form indices	Shape	No. of forms in the form	Nature
{0001}	pinacoid	2	open
{101̄0}, {112̄0}	hexagonal prism	6	open
{hki0}	dihexagonal prism	12	open
{hh2̄hl}	hexagonal bipyramid	12	closed
{hOh̄l} {Ohh̄l}	rhombohedron	6	closed
{hkil}	ditrigonal scalenohedron	12	closed

Since it is sometimes useful to index the faces of trigonal crystals in terms of a rhombohedral cell rather than a hexagonal cell, it is necessary to be able to convert the indices from one type to the other. If the indices referred to the rhombohedral cell are (pqr) and those referred to the hexagonal cell are $(hkil)$ the relationships between them are

$$h = p-q \quad k = q-r \quad i = r-p \quad l = p+q+r$$
$$p = h-i+l \quad q = k-h+l \quad r = i-k+l.$$

Thus it may be noted that the indices of the faces of the unit rhombohedron (parallel to the faces of the rhombohedral unit cell) are obviously (100), (010), (001), ($\bar{1}$00), (0$\bar{1}$0) and (00$\bar{1}$) referred to rhombohedral axes, and these indices transform to those of the form $\{10\bar{1}1\}$ referred to hexagonal axes.

In terms of the hexagonal-type unit cell the trigonal system has one axial ratio c/a, and this can be determined in exactly the same way as in the hexagonal system. In terms of the rhombohedral cell the parameter is the interaxial angle α, and this can be determined in the same way as the interaxial angles of the triclinic system by solving the spherical triangle (100), (010), (001), where the faces are denoted by their rhombohedral indices. Since this triangle has three equal sides all that is needed is the angle between any pair of these faces – which in terms of hexagonal indices are (10$\bar{1}$1), ($\bar{1}$101) and (0$\bar{1}$11). The relationship between c/a of the hexagonal cell and α of the rhombohedral cell is discussed on p. 166.

Problems

1. List the faces belonging to the form $\{210\}$ in the holosymmetric class of each of the following systems: triclinic, monoclinic, orthorhombic, tetragonal, cubic.
2. List the faces belonging to the form $\{21\bar{3}0\}$ in the holosymmetric class of the hexagonal system.
3. Use direction cosines to calculate the angle between the faces (312) and (111) in the cubic system.
4. Express the indices of each of the faces of the form $\{10\bar{1}2\}$ of the holosymmetric class of the trigonal system in terms of rhombohedral indices.
5. Crystals of gypsum are monoclinic with the following interfacial angles:

$$(110):(001) = 82° 19\tfrac{1}{2}',$$
$$(110):(010) = 55° 45',$$
$$(001):(\bar{1}03) = 11° 29'.$$

   Calculate β and $a:b:c$.
6. A tetragonal crystal develops a tetragonal prism form $\{A\}$; a second tetragonal prism $\{B\}$ whose faces make $45°$ angles with $\{A\}$; the basal pinacoid $\{C\}$ and a tetragonal bipyramid form $\{D\}$. The face (D) lies in a zone between (A) and (C) and the angle $A\widehat{}D = 37°$.
   (i) Assume form $\{A\}$ to be $\{100\}$ and make the simplest assumption about the indices of (D). Index the crystal and determine the axial ratio.
   (ii) Assume form $\{B\}$ to be $\{100\}$, again making the simplest consistent assumption about (D). Index the crystal and determine the axial ratio.

CHAPTER 7

The Thirty-two Crystal Classes

CRYSTALS in any particular crystal system all have lattices characterised by unit cells whose shape possesses the same symmetry, but they do not necessarily possess an overall symmetry that is as high as this. The two two-dimensional patterns in Fig. 2.1(a) and Fig. 3.6 have identical unit cells—parallelograms that have 2-fold rotation symmetry. Nevertheless the two patterns as a whole clearly have different symmetry because of the different symmetry of their motifs; the one based on N as the motif has 2-fold symmetry overall, whereas that based on P does not.

In defining the lattice of a pattern we saw in Chapter 2 that an initial point may be taken anywhere in the pattern, and then corresponding similarly situated points throughout the pattern define the lattice regardless of the position chosen for the initial point. If we proceed in this arbitrary way and draw in a particular choice of unit cell in the lattice, as in Fig. 7.1, then the contents of that cell will not in general have the 2-fold rotation symmetry that characterises its shape. However, we have seen on p. 25 that there is a particular

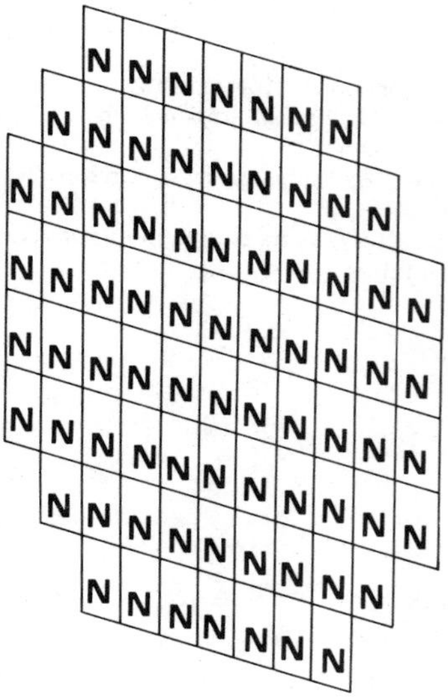

FIG. 7.1. Repeating pattern with motifs having 2-fold rotation symmetry but with cells chosen so that their contents do not have this symmetry.

choice of cell in this pattern that does have contents with 2-fold rotation symmetry. In fact there are four choices for the position of the lattice points relative to the pattern which lead to this result. It occurs if we position the lattice points either at the 2-fold rotation point of the motif or at one of the 2-fold rotation points that occur midway between any pair of adjacent motifs. The resulting unit cells are shown in Fig. 7.2. Thus in the N-pattern it is *possible* to choose a unit cell whose contents have the same symmetry as its shape, whereas it is not possible to do this in the P-pattern. From this point on, when we refer to the symmetry of the contents of a unit cell, the cell will always be understood to have been chosen so that its contents have the maximum symmetry permitted by the nature of the motif. Exactly similar considerations apply in the three-dimensional patterns of crystal structures as in these two-dimensional examples.

FIG. 7.2. The four ways of choosing a unit cell in the pattern of Fig. 7.1 so that its contents have as much symmetry as possible.

It is to be noted that the two patterns in Fig. 2.1(a) and Fig. 3.6 have the same boundaries, so that the external shape of Fig. 2.1(a) would give no hint of its lower symmetry if we were unable to see the internal details of its pattern. Nevertheless the top and bottom boundaries of the P-pattern are different from one another, since one is formed by the heads of the letters and the other by their tails. Such differences in real crystals may lead to an unsymmetrical arrangement of faces that reveals the lack of symmetry of the motif. It is not immediately obvious why this should be so, since if we imagine the pattern of Fig. 2.1(a) to grow by addition of extra motifs at the top and bottom the interactions involved would seem to be the same; in both cases the interaction is between the tail of a P and the head of a P. However, crystal growth is usually a much more complex process than this. Although the attraction of the top and bottom boundaries for appropriately oriented P's would be equal, their attractions for other materials in the environment would be expected to be different. For the pattern to grow, incoming units must compete with other materials adsorbed on the growing face, and this competition will be on different terms at the top and bottom of a crystal like Fig. 2.1(a). The crystal may therefore grow faster at one end than the other, and the face at the *faster growing* end will get *smaller and smaller* and eventually disappear, as shown in Fig. 7.3, provided that there are some inclined faces available to take over from it. However, if the only faces are those parallel to the edges of the unit cell then unequal growth rates will fail to modify the morphology of the crystal. Thus the morphology of a crystal may reveal the true symmetry of the crystal class to which it belongs, or it may correspond to that of a higher symmetry class of the same system. In this chapter we shall first consider the forms that are appropriate to each crystal class, and the varying extents to which they characterise the class, and hen at the end of the chapter consider physical tests that may be applied to determine the class to which a crystal belongs when the morphology leaves it ambiguous.

84   Crystallography for Earth Science Students

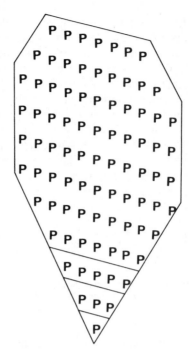

Fig. 7.3. Addition of successive layers to the bottom face of the crystal causes that face to disappear.

Nomenclature

Each crystal class has a symbol which indicates the symmetry elements that are present, using the notation 2, 3, 4 and 6 to denote axes of rotation of the corresponding orders, $\bar{3}, \bar{4}$ and $\bar{6}$ similarly to denote axes of rotation–inversion, and $\bar{1}$ and m to denote a centre of inversion and a mirror plane. The symbol 1 is also used to denote total asymmetry, corresponding to equivalence after a complete rotation about any direction. The symbol of a class not only indicates the symmetry elements present but also their orientations. Up to three non-equivalent directions are defined in a definite order for each system, and any symmetry axes along (and mirror planes perpendicular to) those directions are listed in that order. If there exists both a symmetry axis and a mirror plane perpendicular to it, then both are entered at the same position in the symbol and linked by an oblique stroke, e.g. $2/m$. The conventions for each system are listed below.

Triclinic: The only possibilities here are 1 or $\bar{1}$ so that the question of a conventional order does not arise.

Monoclinic: Only the y-axis can be parallel to the diad so again the question of order does not arise. The possible symbols are 2 or m or $2/m$ depending on whether there is a 2-fold rotation axis alone, a mirror alone, or a mirror perpendicular to a 2-fold axis.

Orthorhombic: The symmetry axis (if any) along, and the mirror (if any) perpendicular to, each crystallographic axis is specified in the order x, y, z. Thus the holosymmetric class is $2/m\ 2/m\ 2/m$.

Tetragonal: The order here is (i) the unique tetragonal axis (z); (ii) the x-axis; and (iii) a direction [110] making 45° with the x-axis. Note that it is unnecessary to specify symmetry along the y-axis since this is necessarily identical to that along the x-axis, because of the tetragonal symmetry.

Cubic: The order here is (i) the x-axis; (ii) the [111] axis, the body-diagonal of the cube, along which there is always a triad; and (iii) a direction [110] making 45° with the x-axis. Again, because of the symmetry, it is unnecessary to specify symmetry elements along y and z.

Hexagonal and trigonal: (i) the unique axis (z); (ii) the x-axis; and (iii) in the hexagonal only, a direction in the xy-plane making 30° with the x-axis.

It is common in crystallographic literature to abbreviate the full symbols of the crystal classes in cases where some of the elements of symmetry are necessarily implied by others. Thus, because mirror-planes at right angles inevitably result in a 2-fold rotation axis along their line of intersection, all the 2-fold axes of the orthorhombic holosymmetric class are inevitably implied by the presence of the mirror planes, and the symbol of this class ($2/m\ 2/m\ 2/m$) is therefore frequently abbreviated to *mmm*. However, this kind of abbreviation tends to confuse the beginner, and full symbols are therefore used throughout this book, with occasional abbreviated symbols in parenthesis to accustom the reader to them.

The thirty-two crystal classes are frequently referred to as the thirty-two *point groups*. This is because the symmetry operations involved leave a point in space unchanged (the origin) and they form a *group* in the mathematical sense. It is by the methods of group theory that it may be proved that the list of thirty-two point groups is exhaustive.

Representation of symmetry and derivation of forms

It is possible to construct three-dimensional models of unit cells containing motifs that exemplify the symmetry of each of the thirty-two point groups, analogously to the two-dimensional unit cells of Figs. 7.2, but such three-dimensional models are difficult to illustrate. It is easier to illustrate solid unit cells decorated on their surface in ways which reduce their symmetry appropriately. Figure 7.50 can be regarded as equivalent to such a set of drawings, and reference may be made to this in order to see how the various point groups are related to unit cell symmetry. However, the most unequivocal way of illustrating the symmetries involved is the stereographic projection of the symmetry elements, and this is given for each crystal class in turn.

In the holosymmetric classes studied in Chapter 6 to exemplify the properties of the crystal systems, the spherical projection (and therefore the stereogram) was divided up by the mirror planes into representative triangles such that any one triangle was repeated by the symmetry to cover the whole sphere. In the lower symmetry classes this concept cannot always be applied without modification, but the same triangles that were defined for the holosymmetric class still enable us to discuss the special and general forms in the same way. The sketch stereograms of the lower symmetry classes are therefore divided into representative triangles by broken lines where mirror planes would exist in the holosymmetric class of the same system, even though any one of these triangles would no longer be repeated by the symmetry to cover the whole sphere. It will be found that special forms often still arise when a pole is placed at a vertex or on a side of such a representative triangle, even when there is no element of symmetry in that position.

It is found that the general form of every class is different and that it always contains fewer faces (by a factor of 2 or 4) than the general form of the holosymmetric class of the same system. The number of faces in special forms is sometimes also reduced relative to that in the holosymmetric class, but not always. These reductions in the numbers of faces in various forms do not mean of course that the omitted faces are impossible; they simply constitute a separate form or forms, related by the symmetry amongst themselves, but not related to the parent face of the original form in the higher symmetry class.

In this chapter we are concerned only with symmetry and not with calculations involving axes of coordinates, and it is therefore most convenient to begin with the triclinic system which has the simplest symmetry, and gradually to work up to the cubic system; however, we shall still treat the hexagonal and trigonal systems last because of their special features and their relationship to one another. In each system we shall begin from the holosymmetric class because we have already studied it in detail, and go downward in symmetry to the less symmetric ones, observing the changes that occur in the forms as various symmetries are eliminated. In according priority to the most symmetric classes in each system we shall indeed only be conforming with nature: more crystals belong to the holosymmetric classes than to any of the others, and indeed there are no certainly known members of one or two of the least symmetric classes.

As far as possible each crystal class is illustrated by a drawing of a mineral crystal whose morphology exhibits the symmetry of the class. It is easy to select such an example for high symmetry classes, but for low symmetry classes the choice has to be restricted to substances which develop a general form, and a good example is not always available, even though relatively common minerals may be known from structural considerations (see Chapter 15) to belong to that class. In such cases non-mineral specimens, or drawings of possible rather than actual morphologies, are used to illustrate the symmetry.

The thirty-two crystal classes fall into three sets, whose significance will become apparent in connection with the physical methods of distinguishing between them (pp. 110–115). The first set comprises those possessing a centre of inversion, the *centro-symmetric* classes, of which there are eleven. All the holosymmetric classes and a few others are centrosymmetric. Among the remaining twenty-one *non-centrosymmetric* classes a further subset of eleven may be recognised which involve no symmetry element involving inversion, i.e. centre of symmetry ($\bar{1}$) mirror plane (m) or $\bar{3}$-, $\bar{4}$- or $\bar{6}$-axis of rotation–inversion. These classes are said to be *enantiomorphous*, and crystals belonging to these may have a right-handed or left-handed character.

The triclinic system

As shown on p. 71 the holosymmetric class $\bar{1}$ of this system has only the general form $\{hkl\}$, which is a pinacoid containing two faces. In class 1 the faces (hkl) and $(\bar{h}\bar{k}\bar{l})$ are unrelated, so in this class the general form contains just one face and is called a *pedion*. If both (hkl) and $(\bar{h}\bar{k}\bar{l})$ are present on a crystal of class 1, they constitute separate forms. The class is obviously enantiomorphous. The stereogram contains a single pole in an arbitrary position. Examples of the two classes are shown in Figs. 7.4 and 7.5.

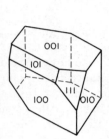

FIG. 7.4. A possible crystal of class 1.

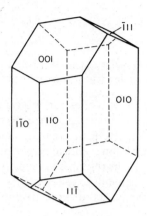

FIG. 7.5. A crystal of class $\bar{1}$ (albite).

The monoclinic system

Figure 7.6 shows the symmetry elements and the general form for each of the classes in this system. In both class m and class 2 the general form contains only two faces which are inclined to one another. If the pole of the face (hkl) of class $2/m$ is considered to be moved to $(h0l)$ on the vertical meridian it will be evident that in class 2 the form $\{h0l\}$ remains a pinacoid while in class m it is not repeated by the symmetry at all and becomes a pedion. Conversely if the pole is moved to (010) the form $\{010\}$ remains a pinacoid in class m but

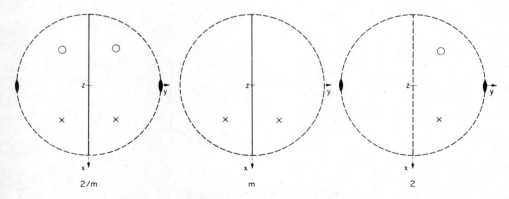

FIG. 7.6. Symmetry and general forms in the three classes of the monoclinic system.

becomes a pedion in class 2. The full comparison of the three classes can therefore be summarised as:

	2/m	m	2
{010}	pinacoid	pinacoid	pedion (also {0$\overline{1}$0} pedion)
{h0l}	pinacoid	pedion	pinacoid
{hkl}	prism	2 intersecting faces	2 intersecting faces
Type of symmetry	centrosymmetric	non-centrosymmetric	enantiomorphous

Special names for the general form in classes m and 2 are avoided here because the conventional ones have become so involved, on the one hand different names being assigned to forms that are of identical appearance but in different orientations, and on the other hand identical names being given to forms of different appearance in the monoclinic and orthorhombic systems because they bear some relation to one another. It would be very difficult to use any of these names in a simplified way without incurring the possibility of further confusion. Examples of the three classes are shown in Figs. 7.7–7.9.

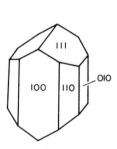

FIG. 7.7. A crystal of class $2/m$ (augite).

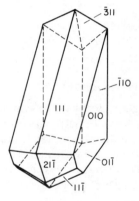

FIG. 7.8. A crystal of class m (hilgardite).

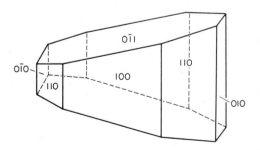

FIG. 7.9. A crystal of class 2 (lactose monohydrate). (From Groth, P., *Chemiscke Krystallographie*, vol. 3, Leipzig, 1910.)

The orthorhombic system

Reference to Fig. 7.10 shows that in class *mm2* one side of the representative triangle becomes a broken line and in class 222 all three sides do. (It is quite arbitrary that class *mm2* is oriented with the 2-fold axis along z: it could equally well be oriented as *2mm* or *m2m*.) Since there is no repetition across the broken line the eight faces of the bipyramid general form $\{hkl\}$ of class $2/m\ 2/m\ 2/m$ (*mmm*) are reduced to four in both cases. In class *mm2* (*mm*) the poles of these four faces all lie in one hemisphere and they constitute a pyramid (an open form), and in class 222 their poles lie in alternate octants and they constitute a *sphenoid*. This is a non-regular tetrahedron whose faces are scalene triangles and whose upper and lower edges are not at right angles (Fig. 7.11);* the name tetrahedron in crystallography is usually reserved for the regular form that occurs in the cubic system.

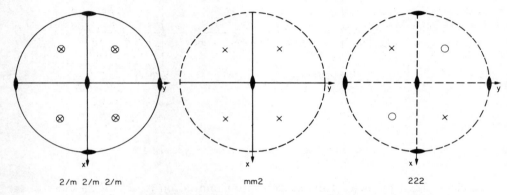

FIG. 7.10. Symmetry and general forms in the three classes of the orthorhombic system.

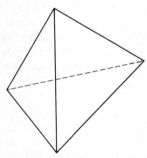

FIG. 7.11. A sphenoid, the general form in class 222. It has four identical scalene triangles as faces, and the top and bottom edges are not at right angles to one another. It is therefore a non-regular tetrahedron.

---

* There is considerable confusion in the literature regarding the name of this figure. Some authors call it a disphenoid or a bisphenoid because they use the term sphenoid to mean a pair of intersecting faces as in the general form of class 2, while others use sphenoid for both that 2-face form and for the non-regular tetrahedron. The usage here is to confine the term to the latter, thereby avoiding ambiguity and at the same time avoiding use of the prefixes. These are undesirable here because in other classes di- indicates repetition of a pair of faces around an axis and bi- indicates repetition of a group of faces by a mirror plane.

90   Crystallography for Earth Science Students

The four faces of the general form of class $2/m\ 2/m\ 2/m$ that are omitted from the general form in the other two classes constitute a second form—in class $mm2$ a second pyramid at the other end of the crystal, and in class 222 a second sphenoid whose poles occupy the other alternate set of quadrants. If a crystal were to exhibit both pyramid forms or both sphenoid forms with the same numerical values of h, k and l it would of course be morphologically indistinguishable from class $2/m\ 2/m\ 2/m$, and this is a general comment on all such cases where a form splits into two in going from higher to lower symmetry.

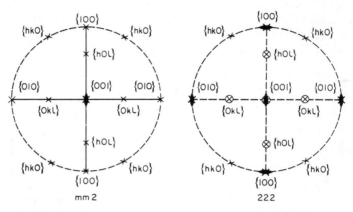

FIG. 7.12. The special forms in classes $mm2$ and 222

In class 222, if the pole of the general face is moved on to any of the sides of the representative triangle (Fig. 7.12) it is still repeated into a prism form (with four poles round the axial great circle on which it lies) just as in class $2/m\ 2/m\ 2/m$, so that the forms $\{hk0\}$, $\{0kl\}$ and $\{h0l\}$ are identical in these two classes, and the same is true of the pinacoids $\{100\}$, $\{010\}$ and $\{001\}$. In class $mm2$, on the other hand, only $\{hk0\}$ remains a prism and only $\{100\}$ and $\{010\}$ remain pinacoids; $\{0kl\}$ and $\{h0l\}$ become just pairs of intersecting faces on one end of the crystal, and $\{001\}$ becomes a pedion.

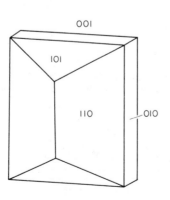

FIG. 7.13. A crystal of class $2/m\ 2/m\ 2/m$ (staurolite).

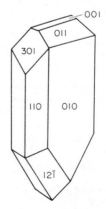

FIG. 7.14. A crystal class $mm2$ (hemimorphite).

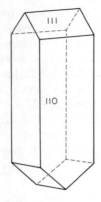

FIG. 7.15. A crystal of class 222 (epsomite).

Examples of the three classes are shown in Figs. 7.13–7.15 and the possible forms are summarised below.

	2/m 2/m 2/m	mm2	222
{100}, {010}	pinacoid	pinacoid	pinacoid
{001}	pinacoid	pedion (also {00$\bar{1}$} pedion)	pinacoid
{hk0}	4-face prism	4-face prism	4-face prism
{0kl}, {h0l}	4-face prism	2 inclined faces	4-face prism
{hkl}	8-face bipyramid	4-face pyramid	sphenoid
Type of symmetry	centrosymmetric	non-centrosymmetric	enantiomophous

The tetragonal system

Discussion of the tetragonal system is made more complicated by the fact that it contains no less than seven crystal classes, six beside the holosymmetric one. In order to keep the treatment closely comparative these are split into three pairs, each of which is compared separately with the holosymmetric class 4/m 2/m 2/m.

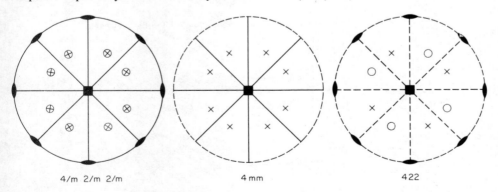

FIG. 7.16. Symmetry and general forms in classes 4/m 2/m 2/m, 4mm and 422.

The first two for comparison are classes **4mm** and **422**, since their relationships to $4/m\ 2/m\ 2/m$ ($4/mmm$) are very similar to those between the classes with analogous symbols in the orthorhombic system, as may be seen by a comparison between Figs. 7.10 and 7.16. As in *mm*2, loss of the equatorial mirror plane in class **4mm** confines the faces of any given form to one end of the crystal unless they lie on the equator, thus leaving prism forms unaffected but converting bipyramids to pyramids and the basal pinacoid to pedions. Similarly, as in 222, loss of all the mirrors in 422 leads to the poles of the general form occupying a set of representative triangles that alternates in all directions with an unoccupied set, and again if the pole is moved on to an edge or vertex of the representative triangle the corresponding form is identical with that in class $4/m\ 2/m\ 2/m$. Examples of these three classes are shown in Figs. 7.17–7.19.

	$4/m\ 2/m\ 2/m$	**4mm**	422
{001}	pinacoid	pedion also {00$\bar{1}$} pedion	pinacoid
{100}, {110}	tetragonal prism	tetragonal prism	tetragonal prism
{*hk*0}	ditetragonal prism	ditetragonal prism	ditetragonal prism
{*h0l*}, {*hhl*}	tetragonal bipyramid	tetragonal pyramid	tetragonal bipyramid
{*hkl*}	ditetragonal bipyramid	ditetragonal pyramid	tetragonal trapezohedron
Type of symmetry	centrosymmetric	non-centrosymmetric	enantiomorphous

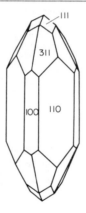

FIG. 7.17. A crystal of class $4/m\ 2/m\ 2/m$ (zircon).

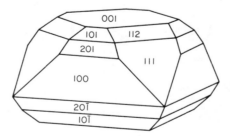

FIG. 7.18. A crystal of class **4mm** (diabolite, $2Pb(OH)_2 \cdot CuCl_2$).

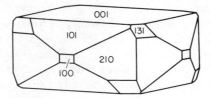

FIG. 7.19. A crystal of class 422 (phosgenite). (Adapted from *An Introduction to Crystallography* by F. C. Phillips, 4th ed., Longmans, 1971.)

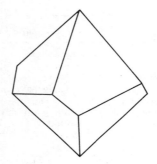

FIG. 7.20. A tetragonal trapezohedron.

The general form of class 422, the tetragonal trapezohedron shown in Fig. 7.20, is like a bipyramid but with the upper and lower halves twisted with respect to each other.

In spite of the close similarity in the relationships between these three classes and those of the orthorhombic system, both in respect of their forms and of their symbols, it should be remembered that the third position in the symbol has a different meaning in the two systems: in the orthorhombic it relates to the third crystallographic axis whereas in the tetragonal it relates to the [110] direction at 45° to the x-axis.

The next pair of classes, $4/m$ and 4, are compared with the holosymmetric class in Fig. 7.21. In class $4/m$ the effect of the removal of the mirrors intersecting in the z-axis, and of

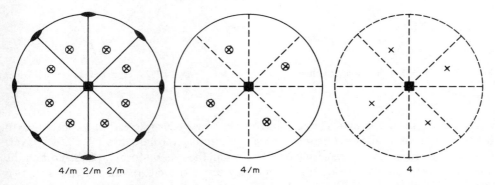

FIG. 7.21. Comparison of the symmetry and general forms of classes $4/m$ and 4 with class $4/m2/m2/m$.

the 2-fold rotation axes perpendicular to it, is to eliminate the difference between the tetragonal and ditetragonal forms, all of which become tetragonal; only alternate representative triangles are occupied around the z-axis, but occupied ones in the upper hemisphere are adjacent to those in the lower hemisphere. It follows that $\{100\}$ and $\{110\}$ cease to be classed as special forms, since they become merely particular and undistinguished cases of $\{hk0\}$. Similar remarks apply to $\{h0l\}$ and $\{hhl\}$ which become particular cases of the general form.

Class 4 is related to class $4/m$ in a similar way to that in which $4mm$ was related to the holosymmetric class by loss of an equatorial mirror, so that all the forms are confined to one end or the other of the crystal.

	$4/m\ 2/m\ 2/m$	$4/m$	4
$\{001\}$	pinacoid	pinacoid	pedion (also $\{00\bar{1}\}$ pedion)
$\{100\}, \{110\}$	tetragonal prism	—	—
$\{hk0\}$	ditetragonal prism	tetragonal prism	tetragonal prism
$\{h0l\}, \{hhl\}$	tetragonal bipyramid	—	—
$\{hkl\}$	ditetragonal bipyramid	tetragonal bipyramid	tetragonal pyramid
Type of symmetry	centrosymmetric	centrosymmetric	enantiomorphous

Examples of classes $4/m$ and 4 are shown in Figs. 7.22 and 7.23.

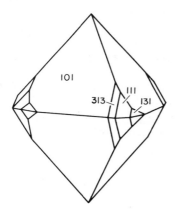

FIG. 7.22. A crystal of class $4/m$ (scheelite).

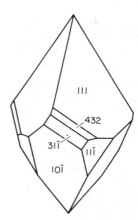

FIG. 7.23. A crystal of class 4 (wulfenite).

The third pair of classes, $\bar{4}2m$ and $\bar{4}$, introduces a new principle with the appearance of the 4-fold axis of rotation–inversion, whose operation is to rotate through 90° *and* invert through the origin. As may be seen from Fig. 7.24, in the general form in class $\bar{4}2m$ half the representative triangles are occupied, but they are disposed in adjacent pairs in two alternate octants in the upper hemisphere and in the interventing two octants in the lower hemisphere. This means that instead of ditetragonal bipyramids we get ditetragonal

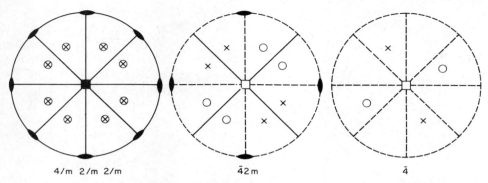

FIG. 7.24. Comparison of the symmetry and general forms of classes $\bar{4}2m$ and $\bar{4}$ with class $4/m\ 2/m\ 2/m$.

sphenoids* for the general form (Fig. 7.25(a)). If the face is moved on to the side of the representative triangle that still contains a mirror plane, it is not repeated by it and occurs only twice in the upper and twice in the lower hemisphere, giving a tetragonal sphenoid (Fig. 7.25(b)) for the form $\{hhl\}$. Unlike Fig. 7.11, this sphenoid has its upper and lower edges at right angles. However, the form $\{h0l\}$ remains a bipyramid since the diagonal mirror plane repeats $(h0l)$ to $(0hl)$ and the 2-fold axis repeats it to $(h0\bar{l})$. All the prisms and the pinacoid remain unchanged.

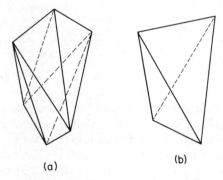

FIG. 7.25. (a) A ditetragonal sphenoid. (b) A tetragonal sphenoid. In (b) the upper and lower edges are at right angles to one another and the faces are isoceles triangles. In (a) the planes of the upper and lower pairs of edges are similarly at right angles.

The relationship of class $\bar{4}$ to $\bar{4}2m$ is rather like that of class $4/m$ to $4/m\ 2/m\ 2/m$ since it involves loss of mirror planes intersecting in the z-axis, and 2-fold rotation axes perpendicular to it, and accordingly the ditetragonal forms become tetragonal, and the

---

* Some authors call these tetragonal bisphenoids, but it keeps the nomenclature more uniform and logical if the duplication is indicated in the adjective; in the ditetragonal sphenoid pairs of faces are repeated by the $\bar{4}$-axis, in place of the single faces of the tetragonal sphenoid, so that the prefix di- has the same significance as in other ditetragonal forms. Some authors use the term ditetragonal scalenohedron by analogy with the corresponding form in the trigonal system, but this conceals the parallelism amongst the tetragonal classes themselves.

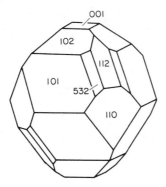

FIG. 7.26. A crystal of class $\bar{4}2m$ (chalcopyrite).

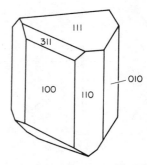

FIG. 7.27. A crystal of class $\bar{4}$ (cahnite).

previous tetragonal forms become just special cases instead of distinct forms, but no other changes occur. Examples are shown in Figs. 7.26 and 7.27.

	$4/m\ 2/m\ 2/m$	$\bar{4}2m$	$\bar{4}$
$\{001\}$	pinacoid	pinacoid	pinacoid
$\{100\}, \{110\}$	tetragonal prism	tetragonal prism	—
$\{hk0\}$	ditetragonal prism	ditetragonal prism	tetragonal prism
$\{h0l\}$ $\{hhl\}$	tetragonal bipyramid	$\left\{\begin{array}{l}\text{tetragonal bipyramid}\\ \text{tetragonal sphenoid}\end{array}\right.$	—
$\{hkl\}$	ditetragonal bipyramid	ditetragonal sphenoid	tetragonal sphenoid
Type of symmetry	centrosymmetric	non-centrosymmetric	non-centrosymmetric

The cubic system

In the cubic system there are five crystal classes. The stereograms of their symmetry and general forms are shown in Fig. 7.28. From the different patterns made by the poles of the faces it is evident that the general form is different in every class. In the present discussion it is worthwhile to make the distinction as to whether the four triads are 3-fold axes of rotational symmetry (3) or 3-fold axes of rotation–inversion ($\bar{3}$), since the two classes that have $\bar{3}$-axes possess a centre of symmetry and the others do not.

In class 432 and class $2/m\ \bar{3}$ the occupied representative triangles alternate with unoccupied ones round each triad, but the relative positions in adjacent octants are related by 2-fold axes and m-planes respectively. In classes $\bar{4}3m$ and 23 alternate octants are empty; in the former every triangle in an occupied octant is occupied, but in the latter only alternate triangles are occupied. As we have seen in Chapter 6, the general form has forty-eight faces in the holosymmetric class $4/m\ \bar{3}\ m$ (abbreviated notation $m3m$); in the class 23, which has the lowest symmetry, there are twelve faces in the general form, and in all the other classes there are twenty-four faces, but arranged differently in each case. These forms rarely, if ever, occur alone on a crystal, and the details of their complicated geometry are therefore of little interest. When they occur as minor forms in combination with others the shapes of their faces are of course dependent on the combination and the relative degree of

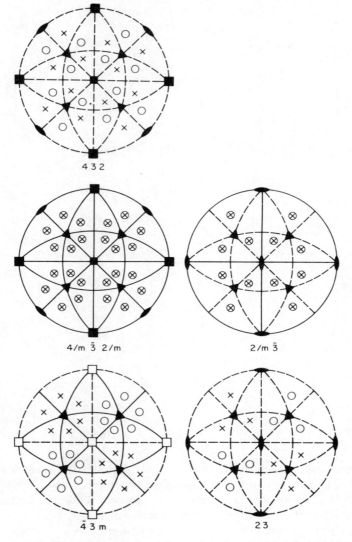

FIG. 7.28. The symmetry and general forms of the five classes of the cubic system.

development, but they can be readily distinguished by their symmetry. A combination of the general form of $4/m\,\bar{3}\,2/m$ with the cube was shown in Fig. 6.14(a), and similar examples are shown for the other five classes in Fig. 7.29. The traditional names of the forms are almost as complicated as their geometry and are more indicative of the shapes of the isolated forms than of their symmetry; these names are therefore omitted, and in the table on p. 101 they are merely indicated as n-hedra and distinguished by a number in parenthesis to show that they differ from other n-hedra with the same value of n.

98 Crystallography for Earth Science Students

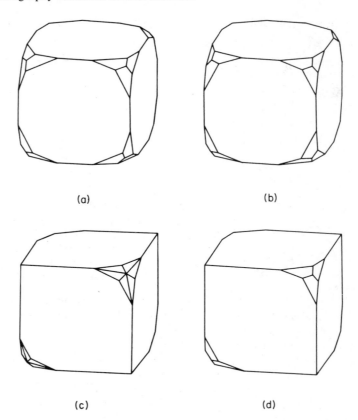

FIG. 7.29. Combinations of the general form with the cube for the four lower symmetry classes of the cubic system: (a) 432, (b) $2/m\bar{3}$, (c) $\bar{4}3m$, (d) 23.

If the pole of a face is moved to (100) in any of the five classes, it may readily be seen from Fig. 7.28 that all poles will converge on those of the cube faces so that this is the form {100} in every class. In the same way {110} remains the rhombic dodecahedron in every class. However, the remaining special forms vary in their behaviour.

Class 432 (abbreviated notation 43) is related to $4/m\,\bar{3}\,2/m$ by loss of all the mirror planes. Nevertheless the remaining rotational symmetry is so high that all the special forms are identical with those of the holosymmetric class that have already been described in chapter 6.

The stereograms of classes $4/m\,\bar{3}\,2/m$, $2/m\,\bar{3}$, $\bar{4}3m$ and 23 have been arranged in a square array in Fig. 7.28 to emphasise their interrelationships: movement from left to right corresponds to loss of diagonal mirror planes, and movement down the page corresponds to loss of axial mirror planes, and each of these losses leads to a specific change in the forms.

It may readily be seen from Fig. 7.28 that if the poles around the triad in the positive octant are made to converge on it, then in classes $4/m\,\bar{3}\,2/m$ and $2/m\,\bar{3}$ poles will converge on the triads in every octant to give the octahedron, whereas in classes $\bar{4}3m$ and 23 they will do so only in alternate octants to give the tetrahedron. Thus in the latter two classes the form $\{111\}$ reduces to a tetrahedron instead of an octahedron. There is then of course a second tetrahedral form $\{1\bar{1}1\}$. The tetrahedral form is rare on its own but is easily recognised in combination with the cube as in Fig. 7.30, when it truncates alternate vertices.

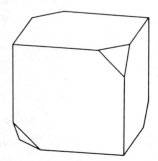

FIG. 7.30. Combination of the tetrahedron with the cube.

When we turn to consider the forms whose poles lie on the sides of the representative triangles we find that if the poles lie on the diagonal planes the form is modified by loss of the axial mirror planes, and if the poles lie on the axial planes then the form is modified by the loss of the diagonal mirror planes, and not vice versa. Thus in class $\bar{4}3m$ the form $\{hk0\}$ is identical with that in the holosymmetric class, namely the tetrahexahedron, like a cube with a low pyramid on each face (Fig. 6.11), whereas in classes $2/m\bar{3}$ and 23 only half of these faces remain in the form, and when present in combination with $\{100\}$ they truncate unsymmetrically each of the cube edges, as in Fig. 7.31(a). This form is quite well known on its own (Fig. 7.31), especially in the case of $\{210\}$ on pyrite, on which account it is often known as the pyritohedron. A more descriptive general name is the pentagonal dodecahedron, but it is to be noted that the faces are not regular pentagons, and the form is quite different in symmetry from the regular pentagonal dodecahedron (one of the five regular solids) that is bounded by regular pentagons and has non-crystallographic symmetry, involving 5-fold rotation axes. There is, of course, a second such form with its non-zero indices reversed, symbolised as $\{kh0\}$.

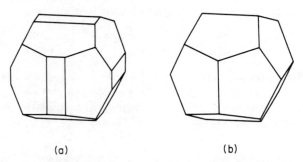

FIG. 7.31. The form $\{210\}$ of classes $2/m\bar{3}$ and 23 (a) in combination with the cube; (b) alone, as in pyrite.

100   Crystallography for Earth Science Students

The remaining two forms $\{hll\}$ and $\{hhl\}$ (both with $h > l$) are the same in class $2/m\bar{3}$ as in class $4/m\bar{3}\,2/m$ (Figs. 6.12 and 6.13), but in classes $\bar{4}3m$ and 23 they have only half as many faces (12), three in each of the alternate occupied octants. Again study of these forms on their own is not very profitable, but combinations of them with the cube are illustrated in Fig. 7.32.

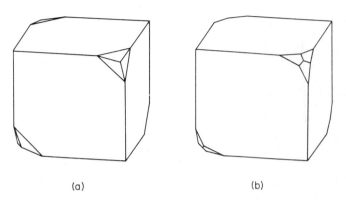

FIG. 7.32. (a) The form $\{211\}$ and (b) the form $\{221\}$ of classes $2/m\bar{3}$ and 23, in combination with the cube.

The forms in the five crystal classes of the cubic system are summarised in Table 7.1.

In Table 7.1 the familiar polyhedral names up to dodecahedron have been retained, different dodecahedra being distinguished by an arbitrary number in parenthesis, as well as by an adjectival description where one is well known. Descriptive form names that were introduced in the holosymmetric system have also been retained, but n-hedron names are added in parenthesis. The distinctions between polyhedra with the same number of faces show which forms permit one to distinguish partly or wholly the class to which a crystal belongs.

The hexagonal system

The seven crystal classes in the hexagonal system are formally analogous to those in the tetragonal system as shown by the similarity of their symbols:

| $6/m\,2/m\,2/m$ | $6mm$ | 622 | $6/m$ | 6 | $\bar{6}m2$* | $\bar{6}$ |
| $4/m\,2/m\,2/m$ | $4mm$ | 422 | $4/m$ | 4 | $\bar{4}2m$* | $\bar{4}$ |

In the first five of the classes in each system (in the order in which they are listed above) the analogy is complete. Stereograms of the symmetry elements and the general form for these five hexagonal classes are shown in Fig. 7.33, which may be compared with Figs. 7.16 and 7.21. There is, of course, an increase in symmetry from 4-fold to 6-fold and corresponding

---

* The difference in the order of the symbols simply arises from the fact that in both systems the face (100) or ($10\bar{1}0$) is taken to be perpendicular to a 2-fold axis, which means that in the tetragonal system the 2-fold axis is along the x-axis whereas in the hexagonal system it is at 30° to the x-axis.

The Thirty-two Crystal Classes

TABLE 7.1

	$4/m\,\bar{3}\,2/m$	432	$2/m\bar{3}$	$\bar{4}3m$	23
{100}	cube	cube	cube	cube	cube
{110}	rhombic dodecahedron (1)	rhombic dodecahedron (1)	rhombic dodecahedron (1)	rhombic dodecahedron (1)	rhombic dodecahedron (1)
{111}	octahedron	octahedron	octahedron	tetrahedron	tetrahedron
{hk0}	tetrahexahedron (24-hedron (1))	tetrahexahedron (24-hedron (1))	pentagonal dodecahedron (2)	tetrahexahedron (24-hedron (1))	pentagonal dodecahedron (2)
{hhl} h>l	trisoctahedron (24-hedron (2))	trisoctahedron (24-hedron (2))	trisoctahedron (24-hedron (2))	dodecahedron (3)	dodecahedron (3)
{hll} h>l	24-hedron (3)	24-hedron (3)	24-hedron (3)	dodecahedron (4)	dodecahedron (4)
{hkl}	48-hedron	24-hedron (4)	24-hedron (5)	24-hedron (6)	dodecahedron (5)
Type of symmetry	centrosymmetric	enantiomorphous	centrosymmetric	non-centrosymmetric	enantiomorphous

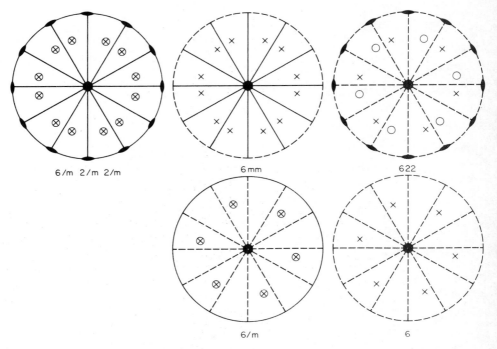

FIG. 7.33. Symmetry and general forms in the classes of the hexagonal system that contain a 6-fold rotation axis.

reduction of the representative triangle from a 45° sector to a 30° sector, but the interrelationships are identical, and the forms in the hexagonal classes may be derived directly from those in the tetragonal ones by substituting the word hexagonal for tetragonal wherever it occurs.

Examples of crystals in these classes are shown in Figs. 7.34 to 7.38.

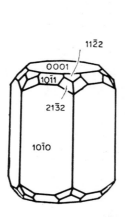

FIG. 7.34. A crystal of class $6/m2/m2/m$ (beryl).

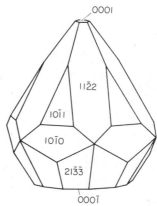

FIG. 7.35. A crystal of class $6mm$ (zincite). (after Palache, C. Amer. Mineral. **26**, 434, Fig. 7, 1941. Copyrighted by the M.S.A.).

TATBLE 7.2

	6/m 2/m 2/m	6mm	622	6/m	6
{0001}	pinacoid	pedion (also {000$\bar{1}$} pedion)	pinacoid	pinacoid	pedion (also {000$\bar{1}$} pedion)
{10$\bar{1}$0}, {11$\bar{2}$0}	hexagonal prism	hexagonal prism	hexagonal prism	hexagonal prism	hexagonal prism
{hki0}	dihexagonal prism	dihexagonal prism	dihexagonal prism	–	–
{h0$\bar{h}$l}, {hh$\bar{2h}$l}	hexagonal bipyramid	hexagonal pyramid	hexagonal bipyramid	hexagonal bipyramid	hexagonal pyramid
{hkil}	dihexagonal bipyramid	dihexagonal pyramid	hexagonal trapezohedron		
Type of symmetry	centrosymmetric	non-centrosymmetric	enantiomorphous	centrosymmetric	enantiomorphous

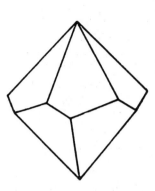

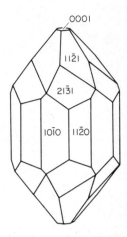

FIG. 7.36. A general form (the hexagonal trapezohedron) in class 622. No good example of morphology is known for this class though the structure of β-quartz belongs to it.

FIG. 7.37. A crystal of class $6/m$ (apatite). (from Schrauf, A., *Atlas der Krystallformen des Mineralreiches*, Vienna, 1877).

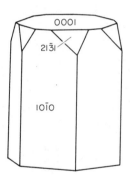

FIG. 7.38. A crystal of class 6 (a possible crystal of nepheline).

The last two classes, involving 6-fold axes of rotation–inversion, differ substantially from the tetragonal classes that involve 4-fold axes of rotation inversion. This is because the action of a $\bar{6}$-axis has rather different results from a $\bar{4}$-axis; the latter, by rotating through 90° and inverting, leaves a point in a quadrant that is at 90° from its starting-point, whereas the $\bar{6}$-axis, by rotating through 60° and inverting, leaves it in a sector 120° instead of 60° from its starting-point. Repetition of the operation leads to six points in all, as in Fig. 7.39, but the arrangement has what looks like trigonal symmetry; it is in fact exactly equivalent to a 3-fold rotation axis with a mirror plane perpendicular to it ($3/m$), but it is regarded as belonging to the hexagonal system by virtue of its representation as $\bar{6}$. It will be seen in the next section that the $\bar{3}$-axis also has somewhat unexpected properties, and these apparent anomalies between the effects of the various axes of rotation–inversion are discussed in an appendix to this chapter.

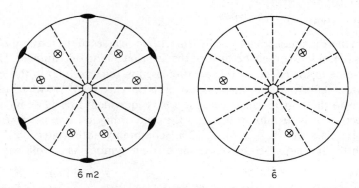

FIG. 7.39. Symmetry and general forms in the classes of the hexagonal system that contain a $\bar{6}$-axis of rotation–inversion.

In class $\bar{6}$ the morphology is very simple. The general form $\{hkil\}$ is clearly a trigonal bipyramid, and if the pole is moved on to the primitive we get a trigonal prism $\{hki0\}$, but the other sides of the representative triangle have no significance and the only other special form is the pinacoid $\{0001\}$ when the pole is on the z-axis. Strangely, the class is purely theoretical and has no certainly known representative in nature.

Class $\bar{6}m2$ is also very rare, but is represented by benitoite, $BaTiSi_3O_9$. Here the sides of the representative triangle retain their significance, and some of the forms have hexagonal symmetry. From Fig. 7.39 the general form $\{hkil\}$ can be seen to be a ditrigonal bipyramid. If the pole is moved on to the mirror plane the form reduces to a trigonal bipyramid $\{h0\bar{h}l\}$ (and there is another similar form $\{0k\bar{k}l\}$), but if the pole is moved on to the side of a triangle midway between the mirror planes the form reduces to the *hexagonal* bipyramid $\{hh\overline{2h}l\}$. If the pole is moved radially from each of these positions on to the primitive then corresponding prism forms are produced as

$$\{hkil\} \rightarrow \{hki0\} \text{ ditrigonal prism;}$$
$$\{h0\bar{h}l\}, \{0k\bar{k}l\} \rightarrow \{10\bar{1}0\}, \{01\bar{1}0\} \text{ trigonal prisms;}$$
$$\{hh\overline{2h}l\} \rightarrow \{11\bar{2}0\} \text{ hexagonal prism.}$$

Finally, of course, if the pole is on the z-axis we obtain the pinacoid $\{0001\}$.

An example compounded from several of these forms is shown in Fig. 7.40.

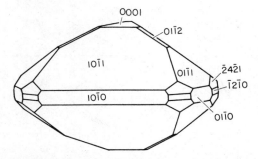

FIG. 7.40. A crystal of class $\bar{6}m2$ (benitoite). (Adapted from *An Introduction to Crystallography* by F. C. Phillips, 4th ed., Longmans, 1971.)

106   Crystallography for Earth Science Students

The trigonal system

Here again there is a measure of parallelism with the classes of the tetragonal system, but it is much less complete, largely because of the fact that 3 is an odd number. Details of the reasons for some of the effects of this are discussed in the appendix to this chapter. The most important effect is that there is no class in this system with a mirror plane perpendicular to the principal axis, because $3/m$ is equivalent to $\bar{6}$ as we have seen in the preceding section. It is because of this that the class of highest symmetry (treated as the holosymmetric class in Chapter 6) is $\bar{3}\,2/m$ (commonly abbreviated $\bar{3}m$). The formal analogy with the tetragonal classes in terms of their symbols is as follows:

		$3m$	32	—	3	$\bar{3}\,2/m$	$\bar{3}$
$4/m\,2/m\,2/m$		$4mm$	422	$4/m$	4	$\bar{4}\,2\,m$	$\bar{4}$

In the trigonal system symmetry elements are listed in only the first two of the usual three positions in the symbol, because a third position is redundant. In the hexagonal and tetragonal systems it denotes a direction at 30° and 45°, respectively, to the x-axis, so that by analogy in the trigonal system it would denote a direction at 60° to the x-axis; but this is just the negative end of the symmetrically related w-axis which is necessarily equivalent to the x-axis by the trigonal symmetry. Thus if the third position were to be used it would always contain the same symbol as the second, and it is in fact better to leave it vacant so as to emphasise the difference from the hexagonal classes, which have the possibility of an additional symmetry element at 30° to the x-axis.

Figure 7.41 shows stereograms of the symmetry elements and the general form in classes $3m$, 32 and 3. Comparison with Figs. 7.16, 7.21 and 7.33 shows a considerable analogy between these three classes and the corresponding ones ($4mm$, 422 and 4) in the tetragonal system and ($6mm$, 622 and 6) in the hexagonal system but it is by no means complete; as we have already seen in Chapter 6 a 30° sector is needed as the representative triangle in the trigonal system to cope with the holosymmetric class, and this is still partly true here. As a result, the analogy that holds between the general forms of these trigonal classes and those of the corresponding tetragonal and hexagonal classes does not extend to all the special forms.

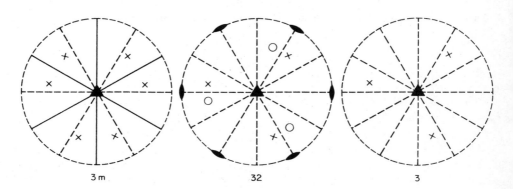

FIG. 7.41. Symmetry and general forms in the classes of the trigonal system that contain a 3-fold rotation axis.

In class 3 no problems arise; just as in classes 4 and 6 the only special forms are {0001} and {hki0}. These are a pedion and a trigonal prism respectively, and the general form is {hkil}, a trigonal pyramid. If the pole is on a side of the representative triangle other than the primitive no special case arises.

In class 3m differences arise because one radial side of the representative triangle lies on a mirror plane and the other one is midway between two such planes. The general form {hkil} is a ditrigonal pyramid, and this reduces to a hexagonal pyramid {hh$\overline{2h}$l} if the pole is midway between mirrors, and to a trigonal pyramid if the pole is on a mirror ({h0$\overline{h}$l} or {0k$\overline{k}$l}). Similarly, when the pole moves on to the primitive the prism forms may be ditrigonal {hki0}, hexagonal {11$\overline{2}$0}, or trigonal ({10$\overline{1}$0} or {01$\overline{1}$0}). The form {0001} is, of course, still a pedion since there is nothing to repeat it.

The situation is similar in class 32, in that the sides of the triangles that join the 3-axis to a 2-axis impart a special character, and so do the sides midway between these. Again in the prism forms there are ditrigonal prisms {hki0}, hexagonal prisms and trigonal prisms, but the last two are related to the indices in the opposite ways: {10$\overline{1}$0} is a hexagonal prism and {11$\overline{2}$0} (and also {2$\overline{11}$0}) is a trigonal prism. The presence of the 2-fold rotation axes leads to a major difference in the forms with $l \neq 0$. Just as in class 622, the general form is a trapezohedron (but a trigonal one of course, Fig. 7.42), and the forms {hh$\overline{2h}$l} and {2h$\overline{hh}$l} are (trigonal) bipyramids; however, the forms {h0$\overline{h}$l} and {0k$\overline{k}$l} are rhombohedra because the faces on the bottom of the crystal are disposed exactly mid-way between those on the top. The form {0001} is, of course, a pinacoid.

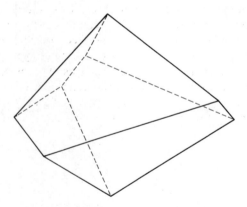

FIG. 7.42. The trigonal trapezohedron, the general form in class 32.

A summary of the forms in these three classes is given below, and examples are shown in Figs. 7.43–7.45.

Figure 7.46 shows stereograms of classes $\overline{3}2/m$ and $\overline{3}$. The former is the holosymmetric class that has been fully described in Chapter 6. It is clear that the general form in class $\overline{3}$ is a rhombohedron, and that the radial edges of the representative triangle have no special significance. When the pole goes on to the primitive one obtains a hexagonal prism {hki0} related to the rhombohedron {hkil} just as in the holosymmetric class the hexagonal prism {10$\overline{1}$0} is related to the rhombohedron {h0$\overline{h}$l}. The form {0001} is a pinacoid because of

108 Crystallography for Earth Science Students

	3m	32	3
{0001}	pedion (also {000$\bar{1}$})	pinacoid	pedion (also {000$\bar{1}$})
{10$\bar{1}$0}	trigonal prism (also {01$\bar{1}$0})	hexagonal prism	—
{11$\bar{2}$0}	hexagonal prism	trigonal prism (also {2$\bar{1}\bar{1}$0})	—
{hki0}	ditrigonal prism	ditrigonal prism	trigonal prism
{h0$\bar{h}$l}	trigonal pyramid (also {0k$\bar{k}$l})	rhombohedron (also {0k$\bar{k}$l})	—
{hh$\bar{2h}$l}	hexagonal pyramid	trigonal bipyramid (also {2h$\bar{h}\bar{h}$l})	—
{hkil}	ditrigonal pyramid	trigonal trapezohedron	trigonal pyramid
Nature of symmetry	non-centrosymmetric	enantiomorphous	enantiomorphous

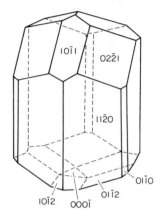

FIG. 7.43. A crystal of class 3m (tourmaline).

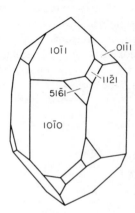

FIG. 7.44. A crystal of class 32 (α-quartz).

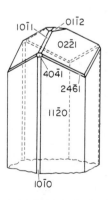

FIG. 7.45. A crystal of class 3 (gratonite). (after Palache, C. & Fisher, J. D. Amer. Mineral. **25**, 257, Fig. 3, 1940. Copyrighted by the M.S.A.).

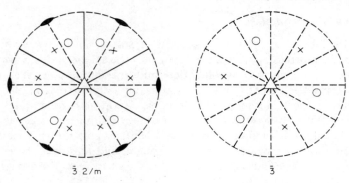

FIG. 7.46. Symmetry and general forms in the classes of the trigonal system that contain a $\bar{3}$ axis of rotation–inversion.

the centrosymmetric character of the $\bar{3}$-axis of rotation inversion. For completeness the forms in these two classes are summarised below, and examples are shown in Figs. 7.47 and 7.48. The ditrigonal scalenohedron has been illustrated in Fig. 6.32.

	$\bar{3}\,2/m$	$\bar{3}$
$\{0001\}$	pinacoid	pinacoid
$\{10\bar{1}0\}$, $\{11\bar{2}0\}$	hexagonal prism	—
$\{hki0\}$	dihexagonal prism	hexagonal prism
$\{h0\bar{h}l\}$, $\{0k\bar{k}l\}$	rhombohedron	—
$\{hh\overline{2h}l\}$	hexagonal bipyramid	—
$\{hkil\}$	ditrigonal scalenohedron	rhombohedron
Nature of symmetry	centrosymmetric	centrosymmetric

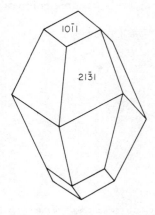

FIG. 7.47. A crystal of class $\bar{3}2/m$ (calcite).

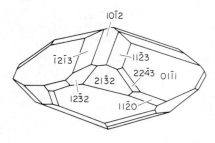

FIG. 7.48. A crystal of class $\bar{3}$ (phenacite).

110   Crystallography for Earth Science Students

Determination of crystal class

If a crystal exhibits a general form, then its crystal class can always be determined uniquely from the morphology, but if it exhibits only special forms this may not always be possible. Indeed it is not always possible to determine the crystal system if only some of the most special forms are present. Thus a crystal having only three pairs of opposite faces at right angles to one another could be cubic $\{100\}$, tetragonal $\{100\}+\{001\}$, or orthorhombic $\{100\}+\{010\}+\{001\}$, and a crystal exhibiting a hexagonal prism plus a basal pinacoid could be hexagonal or trigonal. When such ambiguities arise other sources of information about the symmetry must be sought, and these may be either physical properties or more subtle indications obtainable from the faces themselves.

(i) *Imperfections in the faces*

Up to this point we have assumed that crystals are built up from a perfectly regular repetition of the structural units and accordingly have perfectly smooth faces, but this is not strictly true. Some of the kinds of defect that occur are discussed in Chapter 17, and the existence of defects in the regularity of the structure leads to some crystal faces not being ideally smooth. On one and the same crystal some faces may be smooth and optically bright while others are dull and others may be uneven or striated, but faces that are related to one another by symmetry always have the same sort of defects of this kind. Thus faces that look different are not related by symmetry. This may resolve some kinds of ambiguity; for example, a completely rectangular crystal having one pair of bright faces, one pair of uneven faces and one pair of dull faces would have to be orthorhombic, and not tetragonal or cubic. Similarly if an apparently octahedral crystal has opposite faces bright and dull all round it, then it must be bounded by two tetrahedral forms $\{111\}$ and $\{1\bar{1}1\}$, and must belong to class $\bar{4}3m$ or class 23 of the cubic system, not to a class in which the form $\{111\}$ is an octahedron.

If the imperfections on a face are directional (e.g. striations) their directions must conform with the symmetry emerging on that face, and their directions on different faces must be related by the symmetry of the crystal as a whole. Thus striations on cube faces as in Fig. 7.49 are incompatible with the presence of tetrads and are therefore compatible

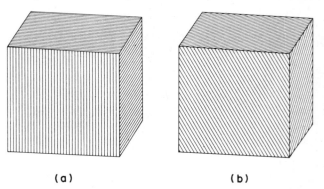

(a)   (b)

FIG. 7.49. Striations on cube faces as shown in (a) are compatible with crystal class $2/m\bar{3}$ or 23, but those shown in (b) are compatible only with class 23.

only with class $2/m\bar{3}$ or class 23. The fact that in Fig. 7.49(a) they are parallel to the cube edges is also consistent with either of these classes; if they were oblique to the edges as in Fig. 7.49(b) this would be incompatible with the presence of the axial mirror planes in $2/m\bar{3}$, but would still be compatible with class 23. This exemplifies an important point in tests of this sort: consistency with a higher symmetry never rules out the possibility of a lower symmetry. In other words, a higher symmetry alternative can never be proved but it can be disproved. However, striations like those shown in Fig. 7.49(a) are often present on cube faces of pyrite which does in fact belong to class $2/m\bar{3}$.

(ii) *Etch pits*

If there are no natural imperfections on the faces of special forms to help in the determination of crystal class, then it is often possible to produce artificial ones by etching the surface with an acid or other suitable solvent that attacks the crystal. Dissolution will usually start at points containing impurities or defects of some kind, and around such a point it will produce a pit bounded by tiny re-entrant faces whose directions may help to define the symmetry. Thus the shapes and orientations of etch pits must accord with the symmetry in just the same way as striations, and they are subject to the same limitations in that they can disprove high symmetry but cannot prove it.

In assessing the implications of etch pits it is, of course, necessary to consider the nature and orientation of any symmetry elements that may be expected to emerge through an etched face, and also the relationship between the orientations of etch pits on different faces. This is helped, especially for faces parallel to the crystallographic axial planes, by the information in Fig. 7.50. This shows crystals of each of the thirty-two crystal classes, bounded by such faces, and decorated on their surface with directional elements, similar to etch pits, sufficient to define the crystal class.

(iii) *Optical activity*

The most straightforward optical investigation of crystals is that which assigns them, on the basis of their effect on polarised light, into one of three categories: isotropic, uniaxial or biaxial. These categories are broader even than the crystal systems, since uniaxial crystals include those belonging to the trigonal, tetragonal and hexagonal systems, and biaxial ones include those belonging to the triclinic, monoclinic and orthorhombic systems. There is, however, one feature of crystal optics which can be used to distinguish between crystals that are enantiomorphous and those that are not, namely rotation of the plane of polarisation of polarised light. To detect this it is necessary to look at an isotropic section of the crystal in the polarising microscope. This is very easy with a cubic crystal, since all sections are isotropic. It is fairly easy with a uniaxial crystal, since it requires a basal section perpendicular to the principal axis. For a biaxial crystal it requires a section perpendicular to one of the two optic axes, and therefore requires a preliminary optical investigation. However, once an appropriate section is available, the observation required is simple; if the crystal is not optically active then no light is visible with the polariser and analyser crossed (at right angles), but if it is optically active then the polariser will have to be rotated from this position before the light is extinguished. Once the extinction condtion is obtained it

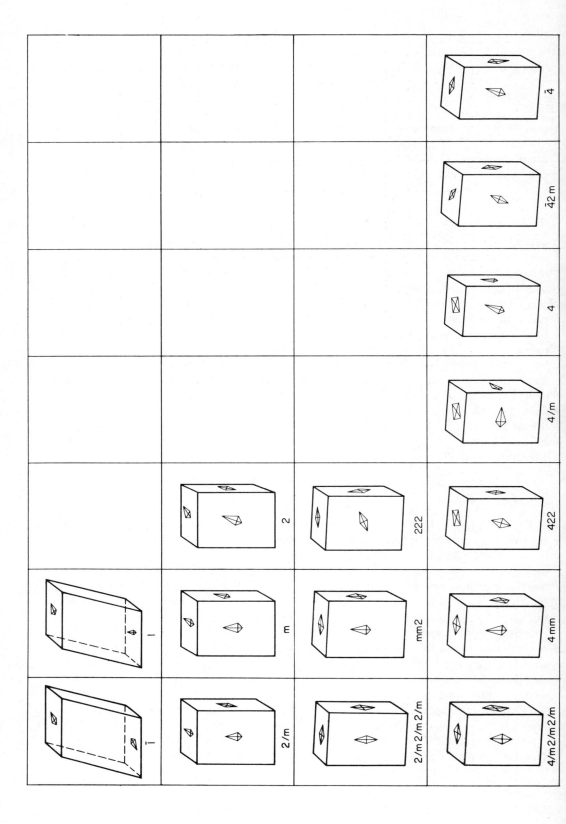

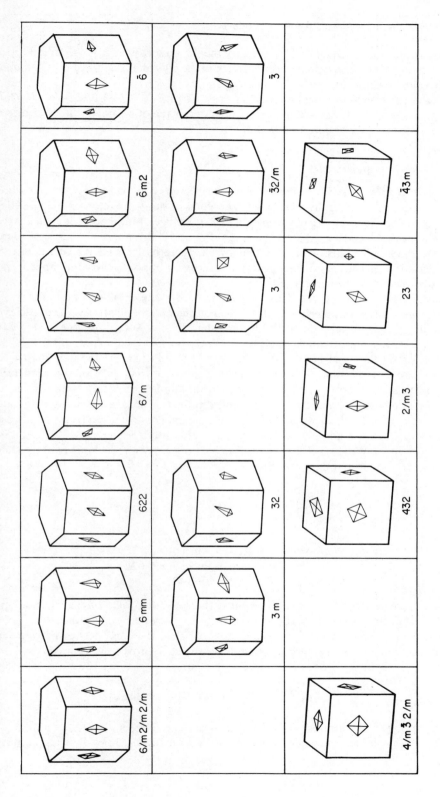

FIG. 7.50. Crystals of the thirty-two crystal classes having axial faces decorated with directional elements similar to etch pits.

will be maintained independently of rotation of the crystal section in its own plane. Optical activity is positive evidence of enantiomorphism in the crystal, though, as usual, failure to detect it is not positive evidence of higher symmetry.

There are eleven enantiomorphous crystal classes, namely 1, 2, 222, 3, 32, 4, 42, 6, 62, 23 and 43; that is, all the classes which possess neither a plane of symmetry, nor a centre of inversion, nor any axis of rotation–inversion.

(iv) *Piezo- and pyro-electricity*

If the crystal does not possess a centre of inversion it may contain polar directions whose opposite ends are structurally dissimilar. The ends of such directions would in ideal circumstances be expected to carry opposite electric charges, but in practice such charges will be neutralised by adsorbed ions. However, if the equilibrium polar charges of such a crystal are modified by distorting it, either by changing its temperature or imposing a mechanical stress, net charges will appear temporarily on opposite ends until appropriate ions can be picked up to neutralise them.

Pyro-electricity is most readily demonstrated in needle-like crystals by dipping them in liquid nitrogen. The resulting charges developed on their ends will attract fine particles of lycopodium powder. Piezo-electricity is best detected by subjecting the material in powdered form to a high-frequency electric field between two metal plates. If the particles are piezo-electric they will be set into forced elastic vibrations by the field, but if the frequency happens to coincide with an elastic resonance frequency of the particle its vibration will become large and absorb a significant amount of energy from the electric field. Thus, if the frequency of the field is changed progressively, then as it passes through particle resonances the absorption of energy can be arranged to cause a feed-back into an audio-frequency circuit and give an audible click on a loudspeaker.

The non-centrosymmetric classes (which include the enantiomorphous ones as a subset) are twenty-one in number, namely:

Triclinic	1
Monoclinic,	2, *m*
Orthorhombic	*mm*2, 222
Trigonal	3, 3*m*, 32
Tetragonal	4, $\bar{4}$, 4*mm*, $\bar{4}$2*m*, 422
Hexagonal	6, $\bar{6}$, 6*mm*, $\bar{6}$*m*2, 622
Cubic	23, $\bar{4}$3*m*, 432

However, the symmetry of class 432 is so high that in spite of its non-centrosymmetric character it cannot exhibit pyro- or piezo-electric properties. In the other non-centrosymmetric classes also it is always possible that these properties may not be detectable, so that once again it is only possible positively to prove low symmetry, and lack of proof of low symmetry does not prove high symmetry.

(v) *X-ray methods*

The most powerful methods of symmetry determination depend on X-ray diffraction, but, as they are complex, evidence from the preceding methods is usually sought first. It used to be thought that X-ray methods could not distinguish between pairs of classes in which one is non-centrosymmetric and the other can be derived from it by adding a centre of inversion. This is true if attention is paid only to the directions of diffracted X-rays. On this basis the triclinic, monoclinic and orthorhombic systems cannot be subdivided and it is only possible to divide the classes in each of the trigonal, tetragonal, hexagonal and cubic systems into two distinguishable sets called Laüe groups. These groupings are:

	Set 1	Set 2
Trigonal	$3, \bar{3}$	$3m, \bar{3}m, 32$
Tetragonal	$4, \bar{4}, 4/m$	$4mm, \bar{4}2m, 422, 4/m\ 2/m\ 2/m$
Hexagonal	$6, \bar{6}, 6/m$	$6mm, \bar{6}m2, 622, 6/m\ 2/m\ 2/m$
Cubic	$23, 2/m\ \bar{3}$	$\bar{4}3m, 432, 4/m\ \bar{3}\ 2/m$

Much more powerful methods are available if the intensities of the diffracted X-ray beams are taken into account, and in favourable cases these can resolve many more ambiguities, but the methods are too complex to be treated in this book.

Appendix on the properties of axes of rotation–inversion

The properties of the $\bar{3}$-, $\bar{4}$- and $\bar{6}$-axes of rotation–inversion seem at first sight to be surprisingly different from one another, and their differences seem to be extraordinarily irregular and confusing. This section is therefore included to help those students who are put off by this apparent lack of system in what is otherwise an extremely systematic subject. It is not, however, essential reading for an understanding of crystallography by those students who are happy to accept the properties of the symmety elements without further enquiry.

The apparent anomalies between the axes of rotation–inversion of different orders arise from the fact that there exist three different kinds of axes of rotation–inversion and only four representatives of these three kinds figure explicitly (and one more implicitly) in crystallography. The three sets differ in the nature of the number specifying their order, according as this is odd, an odd multiple of 2, or an even multiple of 2, i.e. $2n+1$, $2(2n+1)$, or $4n$. The symbols of axes belonging to these sets are therefore:

$$2n+1: \quad \bar{1}, \bar{3}, \bar{5}, \bar{7}, \bar{9}, \bar{11} \ldots$$
$$2(2n+1): \quad \bar{2}, \bar{6}, \bar{10} \ldots$$
$$4n: \quad \bar{4}, \bar{8}, \bar{12} \ldots$$

Of these only $\bar{1}, \bar{2}, \bar{3}, \bar{4}$ and $\bar{6}$ are possible in crystals. Moreover, $\bar{2}$ is not usually discussed as such because it is equivalent to m, and $\bar{1}$ is unique because it is not related to a specific axial direction; thus only one example from each set is usually considered.

In order to depict the effects of the symmetry elements under consideration it is convenient to use stereograms showing the way that a pole is repeated by them, and this is done in Fig. 7.51. It must of course be emphasised that, where the symmetry shown is not crystallographic, these stereograms correspond to artificial faceted forms of the specified symmetries and not to crystals. The first three members of each set are illustrated in order to demonstrate their complete regularity. These stereograms may all be constructed by repeated application of the operation in terms of which the symmetry element is defined: in

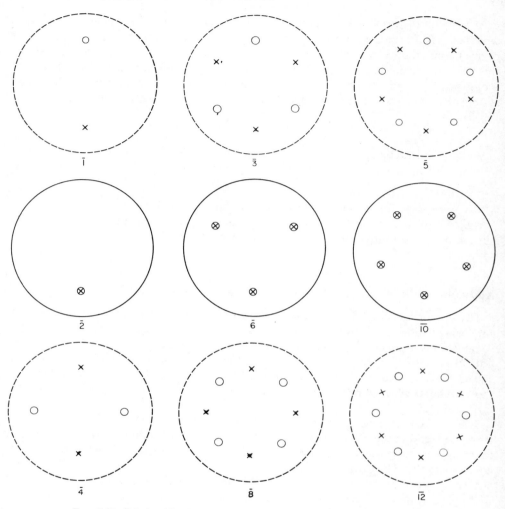

FIG. 7.51. Relationships between axes of rotation–inversion of different orders.

every case a pole is inserted in an arbitrary position; for an $\bar{r}$-axis it is then rotated through $360°/r$ and inverted through the centre, and then its new position is marked in; the same operation is then applied to this new pole, and the process repeated until the original pole is reached.

The first distinction to be noted between the sets is that if r is odd the stereogram contains $2r$ poles whereas if r is even it contains r poles.

The second distinction is that if $r = 2n + 1$ then the $\bar{r}$-axis can be decomposed into an r-fold rotation axis + a centre of inversion. Similarly if $r = 2(2n + 1)$ then the $\bar{r}$-axis can be decomposed into a $\frac{1}{2}r$-fold rotation axis + a mirror plane perpendicular to it, but if $r = 4n$ then $\bar{r}$ cannot be decomposed into simpler elements.

The third distinction is that if the object having an $\bar{r}$-axis is projected on to the plane perpendicular to the axis (equivalent to making no distinction between × and ○ on the stereogram) then the symmetry of the projection is $2r$-fold, $\frac{1}{2}$ r-fold or r-fold according as $r = 2n + 1, 2(2n + 1)$ or $4n$. It is this which leads to the possibility of trigonal prisms in class $\bar{6}$ and hexagonal prisms in class $\bar{3}$, since poles of type × and ○ cease to be distinguishable if they are moved on to the primitive to generate a prism form.

Thus the apparently anomalous properties of $\bar{3}$- and $\bar{6}$-axes relative to $\bar{4}$-axes are entirely regular when viewed in the wider context of general symmetry theory.

Problems

1. To what crystal classes do the crystals belong that are depicted in Figs. 1.3 and 1.5?
2. A crystal has three pairs of opposite faces at right angles to one another. Two pairs are bright and the third pair is striated, with the striations parallel to an edge of the face and in the same direction on the two opposite faces.
   (i) To what crystal classes may the crystal belong?
   (ii) To what crystal classes might the crystal belong if the striations were parallel to an edge but at right angles to those on the opposite face?
   (iii) If the striations were at right angles on the opposite face as in (ii) but were not parallel to any edges, to what class would the crystal belong?
   (iv) What conclusion would you draw if the striations on the two opposite faces were parallel to one another but not parallel to an edge?
3. Use sketch stereograms to derive the indices of all the faces in each of the following forms:
   (i) $\{112\}$ in class 432,
   (ii) $\{112\}$ in class $\bar{4}3m$,
   (iii) $\{210\}$ in class $2/m\bar{3}$,
   (iv) $\{321\}$ in class 23.
4. The phenomenon of pyro-electricity was first observed in the trigonal mineral tourmaline (complex silicate of boron, aluminium, iron, magnesium, etc.). Tourmaline does not show optical activity or enantiomorphous crystal forms. To which crystal class does it belong?

PART II

CHAPTER 8

The Basis of X-ray Crystallography

THE use of X-rays in crystallography is largely unrelated to their property of penetrating through material objects with which they are often primarily associated; it is dependent on the fact that they have a wave-like nature with a wavelength that is of the same order of magnitude as the distances between neighbouring atoms in chemical structures. In fact, as we shall see later, X-rays that have the most useful wavelengths in crystallography have a very restricted ability to penetrate material objects, as compared with the X-rays that are used for medical and industrial purposes. The generation of X-rays suitable for use in crystallography, and their particular properties, will be discussed later in this chapter; in order to introduce their use we need to know merely that they are electromagnetic radiation, like light but with very much shorter wavelength. By suitable experimental arrangements we may obtain X-rays having either a wide range of wavelengths, or a very narrow range (and virtually a specific wavelength), and by analogy with light we describe these as *white X-rays* and *monochromatic X-rays* respectively.

X-rays are *scattered* in all directions when they hit electrons, and they are therefore scattered in their passage through matter by the electrons present in the atoms. However, if the matter concerned is a crystal then the scattering takes place only in specific directions that depend on the repeating pattern of the crystal structure. The direction in which scattering occurs can therefore be used as a means of investigating the nature of the repeating pattern that we were led to postulate in Chapter 2 in order to account for crystal morphology.

A crystal structure is a three-dimensional repeating pattern of atoms in which we can define a lattice of similar, similarly situated, points which define the repetitive character of the pattern. Any three such points define a *lattice plane*, and a set of planes parallel to this will pass through all the points of the lattice and be regularly spaced. The perpendicular spacing between such planes is conventionally denoted as d. A two-dimensional section is shown in Fig. 8.1. Such a set of lattice planes divides up the crystal structure into a set of identical slices of thickness d, and it is convenient to analyse the interaction of X-rays with the crystal in terms of their interaction with such a set of slices.

A single such slice is smooth and flat, and can be shown to reflect an X-ray beam in exactly the way that a mirror reflects light, in that the incident and reflected rays lie in a plane perpendicular to the slice, and the angles of incidence and reflection are equal. To be more precise, it reflects like a very lightly silvered mirror, so that most of the incident radiation is transmitted and a very small fraction is reflected. In the stack of slices that constitute the crystal, there is therefore a weak reflected ray produced from each slice in turn, and these reflected rays interfere with one another destructively unless they have all

122   Crystallography for Earth Science Students

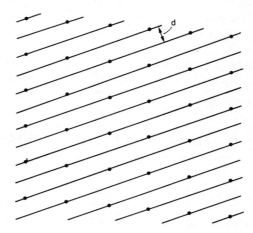

FIG. 8.1. A two dimensional section of a crystal lattice with a particular set of lattice planes marked in. The interplanar spacing is d.

travelled distances that differ by a whole number of wavelengths. From Fig. 8.2 it may be seen that this happens only for the specific values of the angle of incidence (θ) for which $AB + BC$ is a whole number of wavelengths, and which therefore satisfy the *Bragg Equation* $n\lambda = 2d \sin \theta$.

Since the reflected ray arises from the same effective depth in each slice there is no loss in generality from assuming it to arise from the top surfaces of the slices, and Fig. 8.2 is drawn

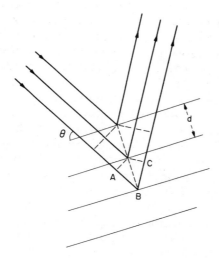

FIG. 8.2. Reflection of X-rays from the first three slices of a crystal defined by some particular set of lattice planes. The extra path length travelled by the waves reflected from each successive layer is given by $AB + AC = 2d \sin \theta$. For constructive interference (reinforcement) this must be a whole number of wavelengths, $n\lambda$.

on this assumption. This is geometrically the same as if the X-rays were reflected by the lattice planes, and many books say that this is what happens. However, that is a "shorthand" description of what really happens, because lattice planes and lattice points are merely mental constructions; X-rays can really be reflected only by material substances, not by mental constructs, and confusion between the two at this stage can lead to difficulties of understanding later on. An initial discussion of the phenomena in terms of slices of crystal structure is therefore much to be preferred.

From the Bragg Equation it is evident that for a given set of slices (with a given value of d) there may be different values of θ depending on the value of the integer n; but the number of these will be limited by the fact that

$$\sin \theta < 1$$

so that

$$n < 2d/\lambda.$$

It follows from this that we are only interested in slices of thickness

$$d > \lambda/2$$

because otherwise there will be no value of n small enough to permit a reflection. On both counts there is therefore a finite number of *Bragg angles* θ for a given wavelength λ.

In order to take the matter further we need to specify in some way the set of lattice planes which we used to slice the crystal. If an origin is taken at a particular lattice point, and a unit cell of the lattice (measuring $a \times b \times c$) is chosen, then axes of coordinates x, y, z, may be defined to lie along these edges of the unit cell. For any set of parallel lattice planes that member of the set that is nearest to the origin will intercept a/h, b/k, c/l from the axis (Fig. 8.3). This particular set of lattice planes can therefore be described by the Miller indices hkl, just like a crystal face. Note, however, that when such indices are used to denote a set of lattice planes they are given without parentheses, to distinguish them from the face (hkl). Also whereas crystal faces are usually restricted to small values of h, k and l, we are often interested in lattice planes with quite high indices, up to 20 or so.

The *interplanar spacing*, d, of a set of lattice planes depends on their indices (as well as on the size of the unit cell) and this fact is often denoted by writing d_{hkl}. From Fig. 8.3 it can be seen that the value of d_{hkl} gets smaller the larger h, k and l.

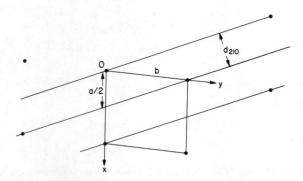

FIG. 8.3. Intersection of a set of lattice planes with the edges of a unit cell having its origin at O. The particular planes illustrated have Miller indices 210.

124   Crystallography for Earth Science Students

In addition to using hkl to symbolise a set of lattice planes, it is also used as a symbol for the X-ray beam "reflected from that set of planes" (to use the shortened description). Strictly speaking it is then necessary to qualify this symbol according to the value of n in the Bragg Equation, and refer to the first order, second order or nth order of the hkl reflection. However, it is much more convenient to describe the nth-order reflection as the reflection nh, nk, nl. Indeed, although the lattice plane (like a crystal face) can only have indices without a common factor, it is convenient to discuss the reflection nh, nk, nl as though it were produced from a set of equally spaced planes whose first member makes these intercepts. Such a plane would clearly be a factor of n closer to the origin than the first hkl plane, so that

$$d_{nh, nk, nl} = d_{hkl}/n.$$

Although most members of the set of planes nh, nk, nl do not in fact pass through lattice points at all, it is convenient to treat them as "honorary lattice planes", because then all reflections can be treated as first order, and the Bragg Equation can be simplified to

$$\lambda = 2d \sin \theta.$$

In this simplified form the equation contains three variables (λ, d and θ) and information about the lattice of a crystal is most easily obtained if λ is kept constant by using monochromatic X-rays; measurements of θ then enable one to determine the d-spacings of lattice planes. The simplest experimental arrangement is to allow a collimated beam of X-rays to strike a specimen of a crystalline material in finely powdered form. Such a specimen contains small crystals in all possible orientations; thus for each set of lattice planes, hkl, there will be many crystals appropriately orientated so that this set of planes will be inclined to the incident beam at an angle θ_{hkl} that satisfies the Bragg Equation for the corresponding value of d_{hkl}. The resulting reflected rays will be deviated by $2\theta_{hkl}$ and will lie on the surface of a cone (in either the forward or backward direction depending on whether $2\theta <$ or $> 90°$) around the incident beam, as shown in Fig. 8.4. If a photographic

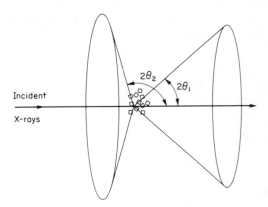

FIG. 8.4. Reflection of X-rays from a powder specimen containing many small crystals in random orientation. Two cones of reflected rays are shown with Bragg angles θ_1 and θ_2, corresponding to two different sets of lattice planes in different crystal fragments. In practice many such cones of rays will be reflected simultaneously.

film is wrapped into a cylinder centred on the specimen it will intercept parts of these cones, which will be recorded after development as lines on the film. From the positions of these lines on such a *powder photograph* the values of θ, and therefore of d, can readily be derived.

The practical details of this and other methods of crystal diffraction and the interpretation of the results obtained are described in the following chapters. The generation and properties of X-rays suitable for such methods are described here.

Figure 8.5 shows the essential features of an X-ray tube. Electrons are produced by a hot filament F and accelerated by a high voltage across an evacuated space to hit a target T, in which X-rays are generated by two processes:

(i) Deceleration of the incident electrons by collision with atoms in the target. An X-ray photon is emitted in each deceleration event, and its wavelength in Å (10^{-10}m) is given by

$$\lambda = \frac{12,398}{\Delta E}$$

where ΔE is the energy (in electron-volts) lost by the electron in the collision. The maximum value of ΔE is the total energy of the electron, which is numerically equal to the

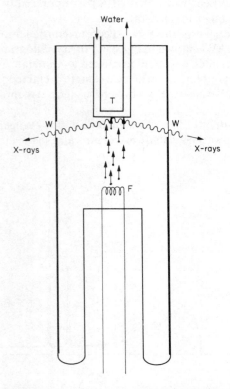

FIG. 8.5. Schematic diagram of X-ray tube. F, filament at high negative potential, emitting electrons; T, water-cooled target at earth potential, from which X-rays are generated; W, W', windows through which X-rays from the target are emitted.

voltage (V) across the tube, and so the minimum wavelength generated is given by

$$\lambda_{min} = \frac{12{,}398}{V}.$$

All wavelengths longer than this will be generated in amounts dependent on the probability distribution of ΔE, which has a maximum at $2V/3$.

(ii) Provided that the voltage exceeds a critical value, ejection of electrons may take place from inner orbitals of atoms in the target, followed by a transition of an outer electron of the same atom into the vacancy so created. This transition is accompanied by emission of an X-ray photon of characteristic wavelength corresponding (by the same equation as above) to the energy change associated with the transition. This is necessarily less than the energy required to eject the electron, so that the wavelength of any monochromatic X-rays generated is always greater than λ_{min}, and is dependent on, and characteristic of, the element of which the target is composed. The most useful characteristic radiation is that emitted following the ejection of an electron from the innermost (K) electron orbital, and is known as K-radiation. It consists of three main lines $K\alpha_1$, $K\alpha_2$ and $K\beta$ depending on the orbital from which an electron falls into the vacancy. The wavelengths of $K\alpha_1$ and $K\alpha_2$ rays are very close together and for many purposes do not need to be distinguished, while that of $K\beta$ is appreciably shorter. All three are necessarily produced together in an X-ray tube.

The distribution of energy over the X-ray spectrum emitted by an X-ray tube with a Cu target and operated at 40 kV is shown in Fig. 8.6. For crystallography the most useful part of the spectrum is the $K\alpha$ lines, and steps are taken to eliminate, or at least minimise the effects of, the rest of the spectrum. The wavelength obtained depends on the target element, and some typical available $K\alpha$ wavelengths (effective mean of $K\alpha_1$ and $K\alpha_2$) are shown in Table 8.1.

When the electron beam hits the target a great deal of heat is generated on a very small area, of the order of a kilowatt per square millimetre, and so the target has to be made of a

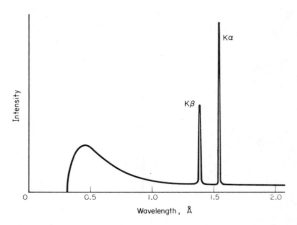

FIG. 8.6. Spectrum of output from a Cu-target X-ray tube operated at 40 kV. The intensity scale is distorted as the true intensity of the characteristic lines is very high and the $K\alpha$ intensity is about 6 times that of $K\beta$. $K\alpha_1$ and $K\alpha_2$ are too close to resolve on this drawing.

TABLE 8.1

	Radiation (λ in Å)		β filter and its K-absorption edge (Å)	
	Kα	Kβ		
Cr	2.291	2.085	V	2.269
Mn	2.103	1.910	Cr	2.070
Fe	1.937	1.757	Mn	1.896
Co	1.790	1.621	Fe	1.743
Cu	1.542	1.392	Ni	1.488
Mo	0.711	0.632	Zr	0.689
Ag	0.561	0.497	Rh	0.534

reasonably refractory metal with good heat conductivity. The Kα wavelength decreases with increasing atomic number of the target, but there are no elements with suitable physical characteristics for the construction of targets to provide wavelengths between those of Cu and Mo. As we have seen, the smallest d-spacing that can give rise to a Bragg reflection is $\lambda/2$; also in any lattice the largest d-spacing is of the order of the edge-length of the unit cell, which is often around 10 Å, and for a d-spacing of 10 Å the Bragg angle is $\sin^{-1}\lambda/20$. For these reasons the number of reflections available with Cr-radiation is usually rather small, while for Mo-radiation the Bragg angles are usually inconveniently small and crowded together, so that an intermediate wavelength such as that from a copper or cobalt target is desirable. MoKα radiation is, however, useful in some circumstances, especially in diffraction studies of single crystals. The effective wavelength used in radiography is that at the peak of the white radiation (0.46 Å at 40 kV, 0.19 Å at 100 kV). Compared with this, the above Kα wavelengths are very long, and are strongly absorbed in most materials, as can be seen from Table 8.2. In general it is evident that absorption increases rapidly both with wavelength and with the atomic number of the absorber. For this reason very low atomic number elements (especially Be) are used for X-ray tube windows to minimise absorption, and lead is used as shielding to avoid radiation hazards. It has to be remembered that although a crystallographic X-ray tube may be intended to produce long wavelength radiation the shielding must be sufficient to reduce

TABLE 8.2
Thicknesses (mm) of various materials to absorb 90% of specified X-radiations

Radiation	Absorber				
	Organic matter (carbohydrate)	Be metal	Al metal	Fe metal	Pb metal
FeKα	1	4	0.09	0.04	0.005
CuKα	2	8	0.24	0.01*	0.009
MoKα	15	42	1.6	0.11	0.017
White peak at 40 kV ($\simeq 0.5$ Å)	46	60	4.5	0.21	0.04
λ_{min} at 40 kV ($\simeq 0.3$ Å)	64	73	17	0.86	0.15

* This value is out of order with respect to wavelength because CuKα radiation is just on the short wavelength side of the K-absorption edge of Fe.

128 Crystallography for Earth Science Students

to negligible levels radiation of wavelength λ_{min} corresponding to the highest applied voltage. Also, although it is easy to provide shielding against them, the long wavelength characteristic radiations can present a greater health hazard than short wavelengths because they are absorbed in the first millimetre or so of the human body and so give high local doses.

If the absorption by a particular chemical element is plotted against wavelength as in Fig. 8.7, it is found that, superimposed on the general tendency to falling absorption with decreasing wavelength, there are specific wavelengths (*absorption edges*) at which the absorption increases abruptly. This happens when the corresponding X-ray photon has just sufficient energy to eject an electron from a particular electronic orbital in an atom, thereby providing an additional mechanism for absorption of the X-ray that is not available at a slightly longer wavelength. The K absorption edge of each element occurs at a wavelength somewhat shorter than that of its Kβ emission line. It is therefore possible to find an element which has its K absorption edge between the Kα and Kβ wavelengths in the output from a given X-ray tube, and this can be used to absorb the Kβ line preferentially. The appropriate element for this *β-filter* always has an atomic number one or two less than that of the X-ray tube target, and appropriate filters and their absorption edges are shown in Table 8.1 for the commonly used radiations. The thickness of the filter is usually chosen to reduce the relative intensity of Kβ:Kα to 1%, and it also leads to a substantial reduction in intensity of the white radiation relative to Kα. The appropriate thickness of Ni-filter for Cu-radiation is 15 μm. Comparative spectral distributions of filtered and unfiltered radiation from a Cu-target tube are shown in Fig. 8.6 and Fig. 8.8.

For most crystallographic purposes filtered radiation is a sufficiently good approximation to monochromatic X-rays. If a more complete elimination of the white component is required for special purposes, this can be achieved by reflecting the rays from a crystal

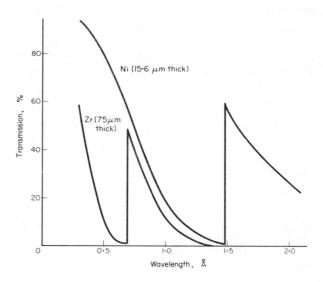

FIG. 8.7. Percentage transmission through two typical β-filters: Ni (for Cu-radiation) and Zr (for Mo-radiation).

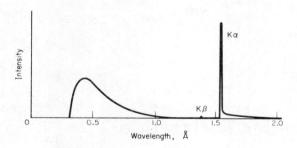

FIG. 8.8. Spectral distribution as in Fig. 8.6 after passage through Ni-filter with absorption characteristics as shown in Fig. 8.7.

plate set at a Bragg angle for the $K\alpha$ wavelength. Indeed, for some particularly critical experiments it is possible to use a device of this type to isolate just the $K\alpha_1$ wavelength alone, eliminating the very close $K\alpha_2$ and all other wavelengths. In the case of Cu radiation these two wavelengths are 1.5505 and 1.5443 Å. But any such process of monochromatisation involves a great loss of intensity, and filtered radiation is adequately monochromatic for many purposes.

Some of the factors involved in the choice of radiation in X-ray crystallography have already been considered. It is fortunate that the $K\alpha$ wavelength of Cu (1.542 Å) is close to the optimum for many purposes in relation to the sizes of the unit cells of crystals, because copper also has optimal properties for construction of the target of an X-ray tube—excellent thermal conductivity and reasonably high melting point. Copper radiation is therefore preferred for general crystallographic purposes unless there is some particular reason to the contrary. One such reason may be the position of the absorption edge of a constituent element of the specimen that is to be studied. If such an absorption edge is just to the long wavelength side of the incident radiation, then this radiation will be very effective in expelling an electron from that element, and thereby exciting it to emit its own characteristic X-radiation in all directions. These *fluorescent* X-rays will then constitute a strong background to the Bragg reflected rays. Such fluorescence arises if $CuK\alpha$ radiation is used to study minerals (or other compounds) containing iron, or to a lesser extent manganese or chromium. However, because of its natural abundance, iron is the main problem, and in the study of iron compounds (Fe absorption edge 1.743 Å) it is often necessary to use a slightly longer wavelength, either $CoK\alpha$ (1.790 Å) or $FeK\alpha$ (1.937 Å). Only the latter is suitable for manganese compounds, as the Mn absorption edge is at 1.896 Å.

Problem

1. If the smallest and largest values of the Bragg angle θ at which reflections can conveniently be recorded on a powder photograph are 4° and 86°, calculate the largest and smallest d-spacings that can be recorded with the following radiations: (1) $CrK\alpha$; (2) $FeK\alpha$; (3) $CuK\alpha$; (4) $MoK\alpha$.

CHAPTER 9

X-ray Powder Diffraction

The Debye–Scherrer powder camera

The most widely used apparatus for photographic recording of X-ray diffraction by a powder specimen is the Debye–Scherrer powder camera. The specimen is usually in the form of a small cylinder about 0.3 mm in diameter, and a few mm long, though only about 1 mm of this length is irradiated by the X-ray beam. If the powder is fine enough to pass a 300-mesh sieve the irradiated volume then contains at least 1000 crystal fragments in random orientations; if the specimen is also rotated about its axis during the experiment each fragment will take up many orientations and each possible set of crystal planes will be enabled to reflect rays around a cone of angle 2θ as in Fig. 8.4. It is not practicable to use a thicker specimen because of the limited penetrating power of X-rays of the appropriate wavelengths. The powder can be filled into a very thin-walled glass capillary tube, since glass gives only a diffuse X-ray scattering pattern which has a negligible effect on the finished photograph. Alternatively the powder may be stuck to the outside of a thin glass rod (about 0.1 mm thick) with some adhesive, or it may be mixed with gum and either rolled or extruded into a cylinder that is self-supporting when dried. The specimen is mounted by means of a small piece of plasticine on to a metal peg that can be rotated and is provided with a screw centring device so that the specimen can be adjusted on to the axis of rotation.

The main body of the camera is a cylindrical drum whose axis is the rotation axis of the specimen holder, which is carried through a bearing in the back of the drum. The strip of photographic film is held against the inside cylindrical surface; one end butts against a fixed peg and the other against a sliding peg (that can be fixed in position by a nut on the outside) to force the film into contact with the camera wall (Fig. 9.1).

In order that we shall be able to measure the angles through which the X-ray beam is deviated by crystal reflection in the specimen it is necessary that the incident beam should be travelling in a well-defined direction. This is effected by means of a *collimator* consisting of two pinhole apertures mounted inside a metal tube (Fig. 9.1) so as to limit the directions of the rays that can reach the specimen. The collimator tube is extended beyond the second aperture towards the specimen in order to intercept rays reflected from crystals in the edge of the second aperture. The tube is wide enough to ensure that it is not hit by the direct beam, but it is as narrow as possible so as to shield as little as possible of the film from rays reflected by the specimen. To this end it is tapered to a conical shape on the outside.

Much of the primary beam passes to each side of the specimen, and some passes through it without being either reflected or absorbed. This undeviated beam passes into the *beam-trap* (Fig. 9.1), a second conically tapered tube, in order to protect the film from diffraction

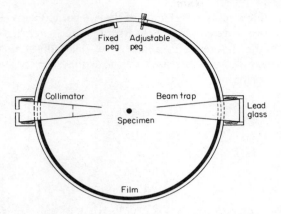

FIG. 9.1. General arrangement of a Debye–Scherrer powder camera.

by air molecules that will occur all the way along its path. Without a beam-trap this air scattering leads to a rather strong background intensity in the low-angle region on the film. The beam-trap extends through the camera wall and terminates in a thick piece of lead-glass to absorb the primary beam. The inside surface of the lead-glass is covered with barium platinocyanide which gives a strong yellow fluorescence when it is irradiated by X-rays, so that a visible indication can be seen through the lead-glass that the camera has been correctly aligned relative to the X-ray tube and that a beam is coming through. This final beam stop is placed a long way down the beam-trap so that the beam-trap tube prevents rays scattered by the beam-stop from reaching the film.

Since both the collimator and beam-trap pass through the cylindrical wall of the camera, two holes are punched in the film to accommodate them. It would be possible to arrange the position of the film in the camera so that either the collimator or the beam-trap passed between the ends of the strip of film, and then only a single hole would be needed at the centre of the strip, and many cameras have been constructed on this principle. However, it is now generally accepted that it is better to have two holes at approximately one-quarter and three-quarters of the way along the film strip, with the ends of the strip located 90° away from the incoming and outgoing beams. With this arrangement the developed film has the appearance shown in Fig. 9.2; the hole through which the beam-trap passed is bracketed symmetrically by lines at low angle ($\theta < 45°$), and the hole through which the collimator passed is bracketed by lines at high angle (θ approaching 90°). The two regions of the film at low and high angles may be distinguished by the appearance of the corresponding lines, because the pattern of lines *either* fades away towards high angles, *or* if the lines remain distinct to high angles then they form doublets. These doublets arise

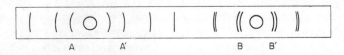

FIG. 9.2. General appearance of a powder photograph obtained from a camera as in Fig. 9.1.

because of the slight difference in wavelength of the two components ($K\alpha_1$ and $K\alpha_2$) of the $K\alpha$ radiation. These two wavelengths are necessarily reflected at slightly different angles. At low values of θ the difference in angle is so small that the two reflections are superimposed, but the difference increases with θ, as may be seen by differentiating the Bragg equation with respect to λ to give

$$1 = 2d \cos \theta \frac{d\theta}{d\lambda}.$$

Thus
$$\frac{d\theta}{d\lambda} = \frac{1}{2d \cos \theta}$$
$$= \frac{\tan \theta}{\lambda}.$$

The sensitivity of θ to λ therefore rises very rapidly when θ approaches $90°$. The $K\alpha_2$ line is always of longer wavelength than the $K\alpha_1$ line and of half its intensity. For copper radiation the wavelengths are:

$$CuK\alpha_1 \quad 1.5405 \text{ Å} \qquad CuK\alpha_2 \quad 1.5443 \text{ Å}.$$

The usually quoted wavelength for $CuK\alpha$ radiation (1.542 Å) is the weighted mean of the two, weighted in terms of their relative intensities.

A film of the type shown in Fig. 9.2 may be measured in one of three ways:

1. For the most accurate work it is fixed to a transparent measuring rule fitted with a vernier cursor, and the positions of all the lines are measured (to an accuracy of 0.1 mm) relative to an arbitrary origin. The mean of the positions of any pair of corresponding low-angle lines (like A and A') gives the position of the incoming beam, the zero of the θ scale; and similarly the mean of two high-angle lines (like B and B') gives the position of $\theta = 90°$. Thus the scale of the photograph in degrees per mm can be found very accurately by taking the averages of many such pairs of measurements. In terms of the scale so found the θ value for any line can then be found since the distance from A to A' is proportional to 4θ (a deviation of 2θ to each side of the incoming beam) and the distance from B to B' is proportional to $360° - 4\theta$.

2. It is usual for the camera to be constructed with a film drum having a radius of 57.3 mm so that 1 mm on the film then subtends $1°$ at the specimen and corresponds to $0.5°$ in θ. This relationship may be disturbed slightly by swelling or shrinking of the film during processing, so that for accurate measurements method 1 must be applied. However, for many purposes it suffices to set the zero of a millimetre scale at the position of the incoming beam (half-way between A and A') and then simply to read off the positions of the lines in mm. The values obtained are then the values of their angles of deviation 2θ.

3. For many purposes the Bragg angles of the lines have to be converted to interplanar spacing, d. Provided that high accuracy is not required it is most convenient to measure the lines using a transparent scale calibrated directly in d-spacings (Fig. 9.3). Like method 2, this method ignores any effect of dimensional change in the film during processing, and since the scale is very non-linear it cannot easily be read as accurately as a millimetre scale. However, such direct reading scales are very convenient and sufficiently accurate for many purposes. It is necessary, of course, to have a different scale appropriately calibrated for each X-ray wavelength that may be used.

FIG. 9.3. A direct reading d-spacing ruler for measurement of powder photographs.

Transmission and back-reflection cameras

Although the Debye–Scherrer camera is the most widely and generally used method of recording X-ray powder photographs, there are a number of other experimental arrangements that are useful. If the specimen is in the form of a thin polycrystalline sheet or film, thin enough to transmit a substantial proportion of the X-rays to be used, its diffraction pattern may be recorded in a transmission camera without breaking up the specimen (Fig. 9.4). A collimated beam is directed perpendicular to the specimen, and the diffraction pattern is recorded on a flat film parallel to the plane of the specimen. Obviously it is only possible by this means to record low-angle reflections out to about $2\theta = 45°$, as beyond that angle they will be absorbed excessively in the specimen, but if high-angle lines are of interest they can also be recorded by a second film, pierced by the collimator. The diffracted rays are recorded as complete circles on the film provided that the specimen contains small crystals in all orientations. If the constituent crystals lie in preferred orientations the diffracted rays may be restricted to limited arcs of such circles, and such a photograph may then be used to study the orientational texture of the specimen. The method is applicable to thin sections (usually of $30\,\mu$m thickness) of very fine-grained rocks, and for this purpose the camera can be made very small with a collimated beam $50\,\mu$m in diameter and a specimen-film distance of about 10 mm. In this way X-ray diffraction patterns can be correlated with features in the rock section that have been studied in the microscope.

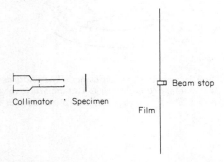

FIG. 9.4. General arrangement of a transmission camera to record a powder photograph of a thin section on a flat film.

Thick massive specimens can also sometimes be studied non-destructively by using the same geometrical arrangement but with the film in the back-reflection position only. This technique finds little application in mineralogical work because most minerals give weak, ill-defined reflections at high angles, but it is applicable to metallurgical specimens because most metals have simple crystal structures that give strong reflections at high angles.

The Gandolfi camera

Although the amount of specimen required for a Debye–Scherrer photograph is only of the order of 0.1 mm^3 or less, this is quite frequently more than is available, or at any rate more than can be conveniently isolated. It may be desired, for instance, to investigate a mineral that occurs in the form of occasional grains less than 0.1 mm in diameter, and the task of assembling some hundred of these to grind into a powder (uncontaminated with anything else) would be prohibitively difficult, and excessively destructive of a rare specimen. In such circumstances the Gandolfi camera can be used. In this camera a single grain is mounted with adhesive on the end of a glass fibre and adjusted to lie in the beam at the centre of a film drum similar to a Debye–Scherrer camera. However, the axis about which the specimen is rotated is at 45° to the beam and its orientation precesses around the axis of the camera at a speed that is not a simple fraction of its speed of rotation (Fig. 9.5). If the specimen grain is regarded as a tiny globe the X-ray beam will therefore be incident on it normally at all positions between latitudes 45°S and 45°N, that is over 71 % of its surface. Thus although the mechanism does not quite achieve the effect of total randomisation of orientation of the specimen, it approaches quite closely to that ideal, and the photographs obtained are almost indistinguishable from powder photographs. Since the specimen is so small a very narrow collimator can be used and the lines of the film are sharper and better resolved than on an average powder photograph. There is a penalty for this, however, in that exposure times are very long, typically 1–2 days instead of 1–2 hours.

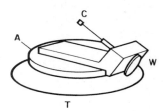

FIG. 9.5. Specimen rotation device in the Gandolfi camera. The platform A is rotated from below, causing the wheel W to roll round the conical track T. This imparts to the crystal incommensurate rotation rates about two axes inclined to one another at 45°.

The Guinier camera

Useful as is the Debye–Scherrer camera (and the derivative forms of it described so far) it has two important disadvantages. The lines tend to be rather broad and there is therefore a lack of resolution of lines lying close together; and also the white continuum in the spectrum of the X-ray tube is reflected over a wide range of angles by each set of planes and leads to a rather dark background to the photographs. This background obscures very weak lines. It is possible to overcome these disadvantages while retaining the basic camera design. Thus, resolution can be improved by using narrower collimators and smaller specimens and the background can be eliminated by reflecting the incident beam from a crystal plate before it enters the collimator so as to isolate the $K\alpha$ radiation and eliminate the continuum. However, either of these procedures leads to an excessive reduction in the intensity of the diffracted rays and therefore to an excessive increase in exposure time; taken together the effect would be prohibitive. The Guinier camera is designed to provide high resolution and low background with reasonable exposure times, and is based on a focusing principle. The incoming X-rays are both focused and monochromatised by reflection (according to the Bragg equation) from a crystal plate that is bent into a curve, as shown in Fig. 9.6. There are two alternative ways of doing this. The original method of Guinier uses a plate that is first ground to follow a circular arc and then plastically bent to an arc of half the radius, the usual crystal used being LiF. The alternative method of de Wolff is to use a thin flat plate of quartz that is elastically bent to follow an arc of an equiangular spiral. In either case a beam diverging from a line source, P (perpendicular to the plane of Fig. 9.6) is made to converge to a line focus, F, and by virtue of the Bragg reflection on the crystal plate it is monochromatic. If a powder specimen, S, is placed in the converging rays on the surface of a cylinder, CC, that passes through F then rays deviated by any particular value of 2θ from all points of the specimen are also focused on the surface of this cylinder. Thus if a film is placed on the surface of this cylinder a powder photograph is recorded on it, and distance along the film is directly proportional to 2θ as on a Debye–Scherrer photograph.

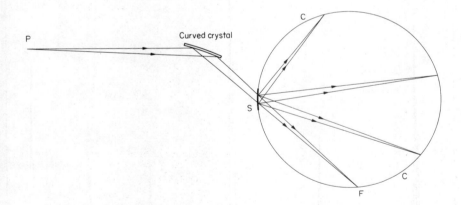

FIG. 9.6. Schematic arrangement of a Guinier focusing camera.

In practice the focusing is not appreciably worsened if the sample lies on a tangent to the film cylinder, and this is easier than curving it. It is stuck on to a thin plastic film such as Sellotape. The upper value of 2θ that can be recorded is maximised if the beam passes asymmetrically across the film cylinder, although even then the range is usually restricted to 0–90°. Within this range, however, a very high-quality photograph is obtained.

136 Crystallography for Earth Science Students

The powder diffractometer

Photographic methods of detecting the diffracted rays have been used in all the experimental arrangements considered so far. This has the merit of being simple, and permits all the accessible reflections to be recorded simultaneously. It does, however, leave much to be desired if one wishes to make a quantitative measurement of the intensity of the diffracted rays. This can be done much better if they are detected electronically in some way. The usual detector is a scintillation counter, consisting of a thin crystal plate (usually NaI with a Tl impurity) which emits a flash of light each time it absorbs an X-ray photon. It is sealed to an electron-multiplier photo-cell which gives out an electrical pulse for each such scintillation. The output of the scintillation counter can then be further amplified, and the rate of arrival of X-ray photons can therefore be recorded on a chart recorder. If the detector can be moved round continuously through a range of Bragg angles the diffraction pattern can be recorded as a sequence of peaks on a chart calibrated in 2θ (Fig. 9.7).

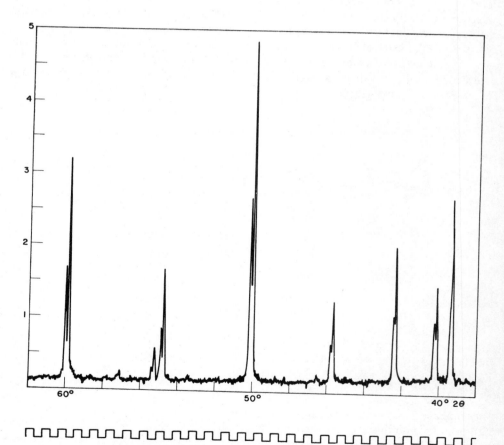

FIG. 9.7. Portion of the output record of a powder diffractometer.

With such an arrangement the various reflections are recorded sequentially instead of simultaneously (as they are on a photographic film), and it is necessary to increase the reflected intensity as much as possible in order to keep the recording time within reasonable limits. This is achieved by using a focusing mechanism somewhat analogous to that in the Guinier camera. The principle is shown in Fig. 9.8(a). X-rays diverge from a line source (P), perpendicular to the page, on to an extended powder specimen (S) which lies in a thin layer on a cylindrical surface that passes through P; this condition means that all the rays make the same angle with the specimen layer, and if this is the Bragg angle θ for a set of planes that lie parallel to the specimen surface the reflected rays will be focused at F, lying on the same cylinder as P and S and in a position symmetrically related to P about S. A narrow slit at F, therefore, passes only rays reflected at the specific angle θ. The detector, D, consisting of a scintillation counter, is mounted immediately behind this slit. Provided that S occupies only a small sector of a cylinder of large radius, the loss of precision of focusing is very slight if S is actually planar. It is then possible to rotate S about its own axis in order

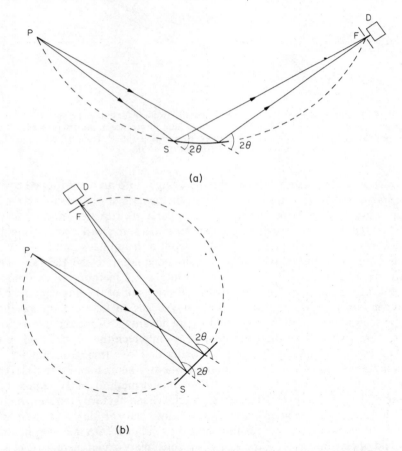

FIG. 9.8. (a) The focusing of rays reflected from a powder specimen in a powder diffractometer. (b) At a higher Bragg angle the focusing circle is smaller but the flat specimen still approximates to it.

to change θ, and if the slit at F, and the detector, are carried on an arm pivoted at the centre of S and rotated twice as fast, the symmetrical focussing condition can be maintained for all values of 2θ from a few degrees to about 160° (Fig. 9.8 (b)). Above this value the slit and detector would in practice get too close to the X-ray source.

In the above description it has been implicitly assumed that the X-rays all travel in planes parallel to the plane of the diagram, while diverging within these planes. In order to prevent divergence out of these planes the rays are made to pass through the spaces between a set of thin parallel metal plates. Such devices are known as Soller slits, and they are placed between the source and the specimen, and also between the specimen and detector system, as shown in Fig. 9.9.

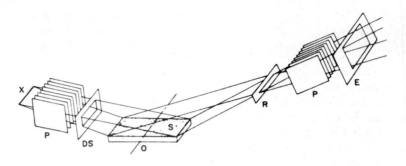

FIG. 9.9. Soller slits permit divergence of the rays parallel to one plane but not perpendicular to it. X, line focus of X-ray tube; P, vertical Soller slits to limit divergence in horizontal plane; DS, divergence slit; S, specimen; O, axis of rotation; R, receiving slit; P, vertical Soller slits; E, antiscatter slit. Distance XO = OR.

The powder specimen may be prepared by packing it into an open-topped cavity in a glass, porcelain or metal plate, the dimensions of the cavity being sufficiently large so that the incident beam does not strike the surrounding parts of the plate. Another convenient way to prepare the specimen is to make the powder into a slurry with acetone; the slurry is then spread on to a piece of microscope slide from a dropping pipette, and when the acetone has evaporated the powder adheres quite adequately to the glass surface.

The diffractometer method is better than photographic methods when quantitative measurements are required of the intensity of reflections or of the intensity distribution across a reflection; it is also quicker if one only needs to measure the positions of reflections over a small range of 2θ angle. It loses this advantage of speed if one requires the whole diffraction pattern, and there are many ways in which the eye can appreciate the qualitative nature of a diffraction photograph better than of a trace from a pen recorder. The two methods are therefore complementary to one another in their usefulness. The film methods tend to be less stringent than the diffractometer method on the requirement for fine grinding of the powder specimen in order to achieve satisfactorily random orientation of the particles, and if the specimen material cleaves into platy or fibrous particles random orientation is not encouraged by spreading these on a flat plate. On the other hand, in the particular field of clay mineralogy, the ease with which platy clay mineral particles orient themselves to lie flat in a diffractometer specimen is turned to good account, and diffractometer studies of such oriented specimens can be more informative than

Debye–Scherrer photographs of more randomly oriented ones. This is because particles lying flat in the specimen give rise only to 00l reflections, and these are the most characteristic and useful reflections from clay minerals.

Uses of powder diffraction

(i) *Identification*

The most widespread use of powder diffraction is for identification of crystalline materials, using their diffraction pattern as a "fingerprint". The simplest way of doing this is by direct comparison of the powder photograph of the specimen with standard photographs of likely materials, and this is practicable if the probable nature of the specimen is fairly well defined in advance from other evidence. For identification to be acceptable all the lines on the photograph must match those of the standard in position and relative intensity. The presence of additional lines on the photograph of the specimen, not present on an otherwise matching standard, will suggest the presence of an impurity which can then be further identified.

If the identity of the specimen is a more open question it can usually be identified by reference to the X-ray Powder Data File published by the Joint Committee on Powder Diffraction Standards (J.C.P.D.S.) which contains the diffraction patterns of some tens of thousands of substances in numerical form, i.e. lists of the d-spacings and relative intensities of their reflections. An example of a card carrying the data for quartz is shown in Fig. 9.10. On this card the intensities of the lines are recorded as I/I_1, relative to the intensity of the strongest line of the pattern which is arbitrarily set at 100. In later additions to the index further information is given as to the intensity of the strongest line itself relative to the most intense line from finely ground and selected corundum, α-Al_2O_3.

5-0490 MINOR CORRECTION

d	3.34	4.26	1.82	4.26	SiO_2					★
I/I_1	100	35	17	35	SILICON IV OXIDE			ALPHA QUARTZ		

Rad. CuKα_1 λ 1.5405	Filter Ni	d Å	I/I_1	hkl	d Å	I/I_1	hkl
Dia. Cut off	Coll.	4.26	35	100	1.228	2	220
I/I_1 G.C. DIFFRACTOMETER	d corr.abs.?	3.343	100	101	1.1997	5	213
Ref. SWANSON AND FUYAT, NBS CIRCULAR 539,VOL. III (1953)		2.458	12	110	1.1973	2	221
		2.282	12	102	1.1838	4	114
Sys. HEXAGONAL	S.G. D_3^4 – P3$_1$21	2.237	6	111	1.1802	4	310
a_o 4.913 b_o c_o 5.405	A C1.10	2.128	9	200	1.1530	2	311
α β γ	Z 3	1.980	6	201	1.1408	<1	204
Ref. IBID.		1.817	17	112	1.1144	<1	303
		1.801	<1	003	1.0816	4	312
8α	n $\omega\beta$ 1.544 $\epsilon\gamma$ 1.553 Sign +	1.672	7	202	1.0636	1	400
2V D_x 2.647 mp Color		1.659	3	103	1.0477	2	105
Ref. IBID.		1.608	<1	210	1.0437	2	401
		1.541	15	211	1.0346	2	214
		1.453	3	113	1.0149	2	223
MINERAL FROM LAKE TOXAWAY, N.C. SPECT. ANAL.:		1.418	<1	300	0.9896	2	402,115
<0.01% AL; <0.001% CA, CU, FE, MG.		1.382	7	212	.9872	2	313
X-RAY PATTERN AT 25°C.		1.375	11	203	.9781	<1	304
		1.372	9	301	.9762	1	320
3-0427, 3-0444		1.288	3	104	.9607	2	321
REPLACES 1-0649, 2-0458, 2-0459, 2-0471, 3-0419,		1.256	4	302	.9285	<1	410

FIG. 9.10. An entry in the J.C.P.D.S. X-ray Powder Data File.

Reference is made to the file via an index* in which the substances are listed in order of the d-spacings of their three strongest lines. The index is divided into sections corresponding to certain fairly narrow ranges of d-spacings of the strongest line (e.g. 3.39–3.32 Å), and within each section substances are listed in order of the d-spacings of their second strongest line. A small section of the mineral index, including the entry for quartz, is shown in Fig. 9.11. Thus, to identify an unknown substance from its diffraction pattern, one has in principle merely to decide which are the three strongest lines of the pattern, measure their d-spacings, and refer to the index, which will lead to the reference number of the substance in the file itself. In practice one will then need to check the d-spacings of the other lines against the values given in the file, or against a standard photograph if one is available. Due allowance has to be made for errors in measurement of the d-spacings and in the estimation of the order of intensities of the lines, especially as the latter are not entirely independent of the experimental method (design of camera or diffractometer, and wavelength of X-rays) which may be different from that used to obtain the pattern in the file.

3.39–3.32

								File No.
$3.34\times$	4.26_4	1.82_2	1.54_2	2.46_1	2.28_1	1.38_1	2.13_1	Quartz, low	5– 490
3.39_9	4.25_7	$2.81\times$	3.97_7	3.12_7	2.59_7	1.72_7	6.51_5	D'Ansite syn	12– 196
3.31_6	$4.24\times$	2.62_8	2.17_6	2.85_5	2.97_4	1.87_3	1.75_3	Rodalquilarite	20– 536
$3.36\times$	4.23_5	1.64_5	2.72_4	2.44_3	2.22_3	1.93_3	3.14_2	Ilsemannite	21– 574
3.31_8	4.23_6	$6.95\times$	3.02_6	2.88_5	2.15_5	1.96_5	1.89_5	Bearsite	15– 378
3.38_9	4.21_7	$4.68\times$	9.39_7	2.54_6	3.79_5	2.45_5	2.37_5	Kratochvilite syn	4– 277
$3.37\times$	4.20_9	$3.46\times$	2.48_9	7.87_8	6.93_5	4.37_4	3.28_4	Bikitaite	14– 168
3.35_9	4.15_8	$4.59\times$	9.30_8	4.24_8	2.59_8	5.10_5	4.78_5	Kratochvilite syn	4– 294
3.38_9	4.13_6	$3.44\times$	2.96_6	2.10_5	4.02_4	3.04_4	1.78_4	Sorbyite	20– 564
3.35_9	$4.08\times$	$2.91\times$	2.57_8	6.71_7	2.74_7	2.52_7	2.18_7	Hungchaoite syn	16– 392
3.40_8	4.04_6	$4.94\times$	2.48_3	7.08_2	4.43_2	2.06_2	1.92_1	Idrialite	18–1727
$3.39\times$	4.04_8	3.92_8	3.03_8	2.75_8	2.67_8	3.79_6	3.66_6	Robinsonite	6– 254
$3.30\times$	4.01_9	2.52_7	2.30_7	2.06_7	5.70_6	4.90_6	3.71_6	Laurionite syn	6– 268
3.34_7	$3.97\times$	3.09_8	2.72_6	3.42_5	2.13_4	6.79_3	5.08_3	Sanbornite	11– 170
3.41_9	$3.84\times$	3.52_9	3.26_9	3.87_7	3.03_7	2.74_3	2.37_3	Mercallite syn	11– 649
3.38_7	$3.81\times$	$6.52\times$	3.23_7	4.33_7	3.26_6	3.01_4	5.91_4	Buddingtonite	17– 517
$3.33\times$	3.79_8	3.22_8	3.29_6	4.24_5	3.46_5	3.00_5	3.26_4	Sanidine, high syn	10– 353
$3.38\times$	3.78_7	2.84_7	3.62_6	2.79_6	4.10_4	2.22_4	2.07_4	Dadsonite	21– 942
$3.31\times$	3.78_7	3.28_6	2.99_6	4.22_6	3.24_5	3.47_5	2.90_3	Orthoclase	22–1212
3.37_5	$3.77\times$	2.78_9	5.85_4	1.99_3	4.93_2	2.92_2	2.47_2	Pabstite syn	18– 196

FIG. 9.11. Part of a page of the index to minerals in the X-ray Powder Data File containing the reference to the entry shown in Fig. 9.10.

(ii) *Quantitative analysis of phases*

A mixture of phases will give a powder pattern containing the lines of each phase, and the intensity of any particular reflection from a given phase will depend on the proportion of that phase present and can therefore be used as a measure of it. In general the relationship is not a simple proportionality because the different phases will absorb the X-rays to different extents, but the method can be calibrated by comparison of the observed powder pattern with patterns obtained from a series of synthetic mixes of known composition.

* If the substance is known to be a mineral the search is greatly simplified by using a very much shorter index containing minerals only. A new file of the cards relating to minerals only has also been published, in book form.

(iii) *Quantitative analysis of solid solutions*

The unit cell dimensions of a mineral change when some element in it is substituted by another. The changes are often approximately, though not necessarily exactly, linear with composition along a solid solution series. If linearity is assumed then the composition of a solid solution can be determined by interpolation between the values for the end-members; otherwise a calibration curve must be derived from measurements on a range of analysed specimens before the composition of an unknown can be inferred from its cell dimensions.

In practice it is not usually necessary in simple cases to determine the cell dimensions themselves, since the d-spacing of some convenient reflection will often serve as a direct indicator of the composition. For example, in the olivine series the substitution of Fe for Mg in forsterite leads to an almost linear variation in d_{130} such that the mole percent content of the forsterite end member (X_{Fo}) is given by

$$X_{Fo} = 4233.91 - 1494.59 d_{130}$$

to within about 4% over the whole range from 0 to 100%.

(iv) *Particle size*

The sizes of the crystallites (i.e. single crystal particles or single crystal domains within a particle) in a powder can affect the diffraction pattern in two ways. If the particles are excessively large there will be too few of them to satisfy the assumption that they lie in all possible orientations. As a result the lines on a powder photograph can be seen to be "spotty", as in Fig. 9.12(a). At the onset of the effect the lines are continuous but contain more intense spots corresponding to rays reflected from the largest particles present. In an extreme case the line may be entirely broken up into spots and there may be only just enough of them for the eye to follow the track of the line. The particle size at which the effect becomes noticeable depends on the size of the collimator (which governs the number of particles irradiated) and on whether the specimen is rotated during the exposure (which increases the number of orientations taken up by each particle). For an ordinary Debye–Scherrer camera with a rotated specimen the onset of the effect is generally taken to occur at a size of about 30 μm, but with a stationary transmission or back-reflection specimens and a fine collimator this may be reduced to about 1 μm. It is possible to use the effect to measure crystallite size by counting the number of spots on a particular line, provided that standards are available for comparison under the same experimental conditions. The effect can also be useful when the constituents of mixtures are being identified, since if the specimen is not too thoroughly ground a hard constituent will give spotty lines when a softer constituent gives continuous lines. This permits one to sort out which lines are due to which constituent, and so facilitates the identification of the constituents with the help of the X-ray Powder Data File.

The observation of spotty lines is only possible by photographic methods. On a diffractometer oversize particles lead to irregular line profiles of which little use can be made.

If crystallites are very small they cause a broadening of the reflections, which can just be detected at dimensions below 1 μm and becomes obvious below 0.1 μm. The reason for this can be understood by reference back to the derivation of the Bragg Equation on the basis

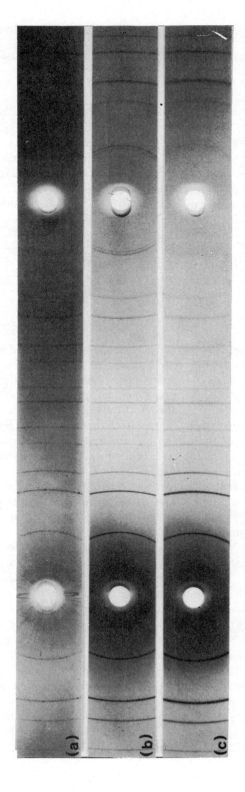

FIG. 9.12. The effect of particle size on powder photographs of fluorite: (a) spotty lines from large particles, (b) normal lines from particles in the 1–30-μm range, (c) broadened lines from very small particles.

of Fig. 8.2. Rays reflected from every slice are in phase, and reinforce one another, when the Bragg angle is satisfied, i.e.

$$2d \sin \theta = n\lambda$$

for integral values of n. When $n = 1.5$ the ray reflected from the top slice is destroyed by interference with that reflected by the second slice down; when $n = 1.25$ it is destroyed by that from the third slice; and when $n = 1 + 1/(2m)$ by that from the $(m+1)$th slice. Thus very large values of m are involved for destructive interference at values of θ close to the Bragg angle, and if the number of lattice planes in the crystal is limited such interference may not occur at all; instead of the reflections being confined to mathematically exact values of the Bragg angle they become broadened to cover a small range of values around the Bragg angle. The angular breadth of the hkl reflection that arises in this way is inversely proportional to the thickness of the crystal measured in a direction perpendicular to the hkl planes. This *particle size broadening* is of course superimposed on the breadth of the reflections that arises from instrumental factors such as the finite size of the specimen in a Debye–Scherrer camera or the imperfect focusing of a diffractometer, and it only becomes obvious when it is of the same order of magnitude. The sensitivity of the powder lines to particle size broadening increases with Bragg angle in proportion to $1/\cos \theta$, so that the broadening first becomes appreciable in its effect on the highest angle reflections, where it leads to a loss of resolution of the $\alpha_1 \alpha_2$ doublets. The effect may be seen in Fig. 9.12(c). For quantitative measurement of particle size by means of this effect the diffractometer obviously has substantial advantages over photographic methods.

(v) *Textures*

If a thin section of material (e.g. a rock) is used as a specimen for diffraction the crystallites in it may have a preferred, rather than a random, orientation. In this case

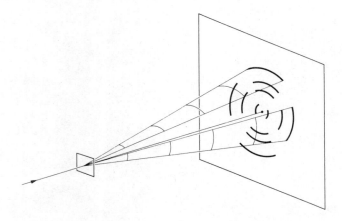

FIG. 9.13. The diffraction pattern obtained by transmission through a thin plate containing small crystallites with a particular kind of preferred orientation texture. The reflections are confined to arcs, instead of the circles that would arise from random orientations.

reflections from some sets of planes may not occur at all, and those that do occur may be restricted to certain parts of the cones of reflected rays (Fig. 8.4). If the diffraction pattern is recorded with a transmission camera one obtains a pattern of arcs instead of circles as shown in Fig. 9.13. From such patterns useful information can be derived about the nature and orientation of the crystallites in the specimen, but the methods of interpretation depend on the theory discussed in subsequent chapters for the interpretation of single crystal photographs, and so cannot be described here.

(vi) *Unit cell dimensions*

The powder pattern of a substance contains in principle the information needed in order to determine the absolute dimensions and interaxial angles of the unit cell. The methods of interpreting diffraction data in these terms are dealt with in the following chapters.

Problems

1. A powder photograph taken with CoKα radiation is measured from an arbitrary origin. Two pairs of lines, A, A', B, B' (as in Fig. 9.2) are at the positions 67.88, 122.62, 258.91 and 290.67 mm. Calculate the *d*-spacings of the two sets of planes that gave these lines.
2. A diffractometer trace of powdered olivine gives the 130 reflection at an angle of deviation of the beam (2θ) of 32.14° with CuKα radiation. What is the mole percentage of the forsterite end-member in the olivine?

CHAPTER 10

Intensities of X-ray Reflections

INSPECTION of any X-ray photograph or diffractometer chart reveals that the reflections corresponding to different sets of lattice planes differ widely from one another in intensity. There are a number of factors responsible for the relative intensities of the reflections from a crystal, most of which involve the specific experimental conditions, but the major one is independent of these and involves only the arrangement of atoms in the crystal structure. This is the reason that powder photographs provide such a good method of identification of crystalline substances, because it is difficult to imagine anything more characteristic of a substance than the arrangement of atoms within it. It is also, of course, the reason why X-ray diffraction enables us to determine the structure of crystals on an atomic scale.

In deriving the Bragg Equation in Chapter 8 we saw that a crystal is divided up by a set of lattice planes into a set of identical slices, and that any one such slice, if it were isolated, would reflect X-rays as a mirror reflects light – except that it would reflect only a very small fraction of the X-ray intensity incident on it, like a very slightly silvered mirror. The intensity of a reflection corresponding to a given set of lattice planes therefore depends on the reflectivity of an isolated member of the set of slices into which those planes divide the crystal. Now the intensity, I, of an electromagnetic wave is proportional to the square of its amplitude, A:

$$I \propto A^2.$$

We therefore evaluate the amplitude of a wave reflected by a crystal slice as the resultant of combining together the X-ray waves scattered in the direction of the reflected ray by atoms at various depths within the slice. First of all, therefore, we must say something of the way in which an isolated atom diffracts X-rays.

The atomic scattering factor

It is the electrons within an atom that are responsible for its scattering of an incident X-ray wave. These electrons are very much concentrated towards the centre of the atom, but they are spread throughout its volume. If an X-ray wave is scattered by the electrons in the direction of its propagation (Fig. 10.1(a)) the scattered waves from electrons at all points within the atom will travel the same total distance (before scattering and after scattering), and they will therefore be all exactly in phase with one another. The total amplitude of the wave scattered by an atom of atomic number Z will be Z *times* the scattering by one electron. Scattering at any other angle, however, will involve differences

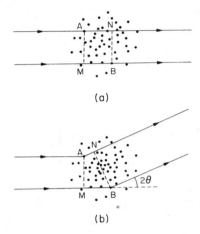

FIG. 10.1. Scattering of X-rays by electrons at two different points (a) in the forward direction, and (b) with a deviation 2θ. The difference in path length travelled by the two rays is in each case MB–AN.

in path length travelled by waves scattered in different parts of the atom (Fig. 10.1(b)) and when the waves reach any given point they will differ in phase. Since the resultant amplitude from the combination of two waves of different phase is always less than the sum of their amplitudes, the scattered amplitude falls off progressively with increasing scattering angle in a manner which depends on the distribution of electron density in the atom, the fall-off being less rapid the more centrally concentrated is the electron density. The difference in path length travelled by two waves depends on $\sin \theta$ (where θ is half the angle of deviation) and their phase difference therefore depends on $\sin \theta / \lambda$. The *scattering factors* (symbolised by f) of all possible atoms have been computed and tabulated as functions of $\sin \theta / \lambda$,* and a few of them are shown graphically in Fig. 10.2. Their

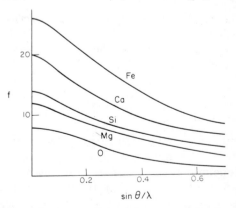

FIG. 10.2. Scattering factors of some elements as a function of $\sin \theta / \lambda$.

* See, for example, *International Tables for X-ray Crystallography*, Kynoch Press, 1962.

numerical values are in terms of the scattering power of a single, classical, isolated electron. It is to be noted that, because f is a function of $\sin\theta/\lambda$, a given atom makes the same contribution to a given Bragg reflection from a given crystal regardless of the wavelength of the X-rays used, since from the Bragg Equation

$$\frac{\sin\theta}{\lambda} = \frac{1}{2d_{hkl}}$$

and it is uniquely defined for a given reflection hkl.

Amplitude of reflection from a crystal slice

Consider a crystal slice, of thickness d, between an hkl plane through the origin of a unit cell and the next plane of that set. Take as the standard of zero phase a wave scattered at Bragg angle θ from something (whether actually present or not) at the top surface of the slice. It follows that the phase of a wave scattered at the bottom of the slice (coincident with the top of the next slice) will be one whole wavelength behind, and that of a wave scattered at any depth pd within the slice will be a fraction p of a wavelength behind the standard phase.

The standard wave of unit amplitude that would be scattered by an electron at the top of the slice may be represented at a particular distance as a function of time, t, as

$$\cos\frac{2\pi ct}{\lambda}$$

where c is the velocity of light. It follows then that a wave scattered by an atom 1 at a fractional depth p_1 in the slice will be represented by

$$f_1 \cos 2\pi\left(\frac{ct}{\lambda} - p_1\right)$$

where f_1 is the scattering factor of atom 1. Similarly the wave from an atom 2 at fractional depth p_2 will be

$$f_2 \cos 2\pi\left(\frac{ct}{\lambda} - p_2\right)$$

and the resultant wave from the two atoms will be

$$f_1 \cos 2\pi\left(\frac{ct}{\lambda} - p_1\right) + f_2 \cos 2\pi\left(\frac{ct}{\lambda} - p_2\right)$$

$$= \left(\cos\frac{2\pi ct}{\lambda}\right)(f_1 \cos 2\pi p_1 + f_2 \cos 2\pi p_2) + \left(\sin\frac{2\pi ct}{\lambda}\right)(f_1 \sin 2\pi p_1 + f_2 \sin 2\pi p_2).$$

In general, if there are many waves adding together this will become

$$\left(\cos\frac{2\pi ct}{\lambda}\right)\sum_i f_i \cos 2\pi p_i + \left(\sin\frac{2\pi ct}{\lambda}\right)\sum_i f_i \sin 2\pi p_i$$

which is a wave of amplitude

$$\left\{\left(\sum_i f_i \cos 2\pi p_i\right)^2 + \left(\sum_i f_i \sin 2\pi p_i\right)^2\right\}^{1/2}.$$

If the structure is centrosymmetric, so that for every atom at p_i there is another similar one at $-p_i$, then the term $\sum_i f_i \sin 2\pi p_i$ will be zero, and the amplitude of the wave simplifies to

$$\sum_i f_i \cos 2\pi p_i.$$

For the sake of the simplification of the mathematics we shall confine our attention to this case, which is very frequently applicable.

We now have to relate the fractional depth (p_i) of an atom in the slice to its coordinates in the unit cell. It is usual in describing crystal structures to use fractional coordinates of the atoms (x, y, z), where these are fractions of the unit cell edges a, b, c rather than actual distances parallel to the axes. The evaluation of p_i is then to be made in terms of the fractional coordinates (x_i, y_i, z_i) and is most easily done in terms of vectors. A two-dimensional analogue is shown in Fig. 10.3. If $\hat{\mathbf{n}}$ is a unit vector perpendicular to a set of planes hkl, and the unit cell edge vectors are $\mathbf{a}, \mathbf{b}, \mathbf{c}$, then the vector $\mathbf{r}$ to any point in the first plane of the set has a projection on the perpendicular from the origin given by

$$\mathbf{r} \cdot \hat{\mathbf{n}} = d. \tag{10.1}$$

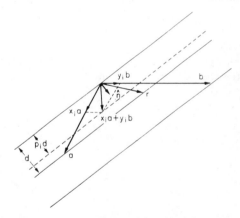

FIG. 10.3. Vector notation for evaluation of the depth ($p_i d$) of an atom at x_i, y_i in a slice of a crystal between two successive $hk0$ planes (drawn with $h = 1, k = 2$).

Thus the vectors which the plane intercepts on the axes all satisfy this equation,

i.e.
$$\frac{\mathbf{a}}{h} \cdot \hat{\mathbf{n}} = \frac{\mathbf{b}}{k} \cdot \hat{\mathbf{n}} = \frac{\mathbf{c}}{l} \cdot \hat{\mathbf{n}} = d. \tag{10.2}$$

Also the ith atom is at the end of a vector $x_i \mathbf{a} + y_i \mathbf{b} + z_i \mathbf{c}$ and lies on the parallel plane

$$\mathbf{r} \cdot \hat{\mathbf{n}} = p_i d. \tag{10.3}$$

It follows that
$$(x_i\mathbf{a} + y_i\mathbf{b} + z_i\mathbf{c}) \cdot \hat{\mathbf{n}} = p_i d \tag{10.4}$$
and we may substitute for $\mathbf{a} \cdot \hat{\mathbf{n}}$, etc., from eqn. (10.2) to give
$$hx_i + ky_i + lz_i = p_i.$$
Thus the amplitude of the wave reflected from the slice is
$$\sum_i f_i \cos 2\pi(hx_i + ky_i + lz_i).$$
Note that this derivation is valid for any system of crystal axes, whether rectangular or oblique.

The summation in this expression obviously has to be taken over all atoms in the slice, but in evaluating the amplitude reflected from the whole crystal the contributions from all the slices will add together since they are all in phase, so that ultimately the summation is taken over all the atoms in the crystal. The result then depends on the size of the crystal, so it is convenient to standardise the calculation in terms of the amplitude reflected per unit cell, and this is called the *structure factor*: F_{hkl} for the particular set of planes hkl. Thus
$$F_{hkl} = \sum_i f_i \cos 2\pi(hx_i + ky_i + lz_i) \tag{10.5}$$
for a centro-symmetric crystal, with the summation taken over all atoms in the unit cell.

It may appear anomalous at first sight that we can take the summation over the whole unit cell when the slice does not extend through the whole of any one unit cell. However, the slice includes parts of many unit cells which together make up a whole unit cell (Fig. 10.4), and the coordinates of atoms (x, y, z) in different unit cells differ only by whole numbers. Since
$$\cos 2\pi(h(x+n)) = \cos 2\pi hx$$
it does not matter whether the atoms added in to the summation are all in the same unit cell or in corresponding parts of different cells.

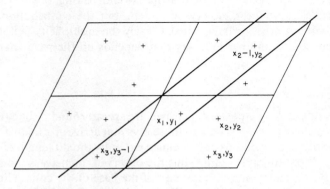

FIG. 10.4. A slice of a crystal between two successive planes does not contain all the atoms in any given unit cell, but those omitted from one cell are included from adjacent cells.

Other factors affecting intensities

Since the intensity of an X-ray beam depends on the square of its amplitude, the intensities of X-ray reflections depend on F^2, but they also depend on a number of other factors.

(i) *The temperature factor*

The above derivation of the structure factor assumes that the atoms are at rest, whereas in fact they are vibrating about their mean positions. If all the atoms are vibrating equally and isotropically it can be shown that this leads to a reduction in the intensities of the reflections by a factor $\exp(-B\sin^2\theta/\lambda^2)$, where B is proportional to the mean square amplitude of oscillation of the atoms, and for a given substance depends on the temperature. It also depends on the rigidity of bonding of the atoms. Thus the diffraction patterns of soft substances like halite show a much more rapid fall in intensity with $\sin\theta/\lambda$ (i.e. along the length of the pattern independently of the radiation used) than do hard substances like quartz or diamond. A more precise analysis requires a different value of B to be associated with each atom so that the temperature factor has to be incorporated in the structure factor formula as

$$F = \sum_i \exp(-B_i \sin^2\theta/\lambda^2) f_i \cos 2\pi(hx_i + ky_i + lz_i).$$

(ii) *The multiplicity factor, M*

The intensity of a reflection from a powder specimen will obviously depend on the number of crystal fragments that are in an appropriate orientation to reflect. This depends in part on the crystal symmetry. The same interplanar spacing, d_{hkl}, will apply to all the sets of planes in a form $\{hkl\}$, and therefore crystals oriented with any member of this form in the Bragg orientation will contribute to the reflection. If the crystal is holosymmetric the intensity due to F_{hkl}^2 will then be multiplied by the number of faces in the form $\{hkl\}$; if it is of lower symmetry there will still be contributions to the powder reflection at the same Bragg angle from all the sets of planes that would belong to the form $\{hkl\}$ in the holosymmetric class of the same crystal system, but the contributions from all these components will not necessarily be equal. Clearly the multiplicity factor varies abruptly from one reflection to another in a way that depends on the particular substance.

(iii) *The arc length factor, l*

The X-rays reflected by a powder specimen are spread round a cone (Fig. 10.5) but the measured intensity corresponds only to those rays that arrive at a small, fixed length of the circumference of the base of the cone – defined either by the slit length of a diffractometer or a unit area of photographic film. Thus the observed intensity is modified by a factor inversely proportional to the circumference of the base of the cone. This factor changes systematically with the Bragg angle and is proportional to $1/\sin 2\theta$. It has a minimum value at $\theta = 45°$ ($2\theta = 90°$), i.e. half-way between the forward and backward scattering

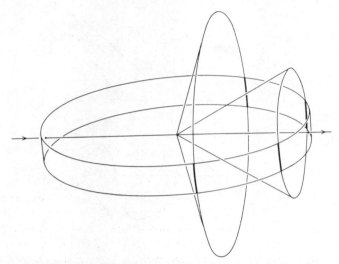

Fig. 10.5. The proportion of a diffracted cone of rays that is intercepted by a given length of line on a powder photograph varies with the angle 2θ.

directions. Whereas the structure factor, the temperature factor, and the multiplicity factor are characteristic of a particular reflection hkl from a particular substance, the arc length factor depends on the X-ray wavelength used. For example, if one compares powder photographs of the same material taken with CuKα and MoKα radiation, high-angle lines on the former ($\theta \sim 80°$) will be at moderate angles on the latter ($\theta \sim 26°$) and will be relatively weakened by a factor of about 0.4, whereas medium-angle lines on the former ($\theta \sim 45°$) will be at low angle on the latter and relatively strengthened by a factor of about 1.6.

(iv) *The polarisation factor, P*

When a plane polarised electromagnetic wave is reflected, the amplitude of the reflected wave is proportional to the component of the incident amplitude in the direction of vibration of the reflected wave. Thus if the incident wave is polarised in a direction perpendicular to the plane of the incident and reflected rays (Fig. 10.6(a)) so is the reflected wave, and the two directions of polarisation are parallel to one another whereas if it is polarised in the plane of reflection so is the reflected wave, and the two polarisation directions are at an angle 2θ (Fig. 10.6(b)). The reflected amplitude A_r is then reduced by a factor $\cos 2\theta$ relative to the incident amplitude A_i. If the incident rays are unpolarised they may be resolved into two components parallel and perpendicular to the plane of reflection, each of intensity $\tfrac{1}{2}$. The combined intensity of the two reflected components is therefore reduced by a factor

$$P = \frac{1 + \cos^2 2\theta}{2}.$$

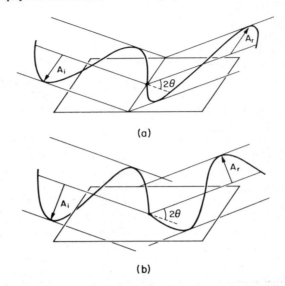

FIG. 10.6. Reflection of a polarised wave from a plane. Plane of polarisation (a) perpendicular; and (b) parallel, to plane of incident and reflected rays.

It is to be noted that the reflected wave is partially polarised, so that if preliminary monochromatisation of the X-ray beam has been effected by reflection from a crystal plate (p. 135) account has to be taken of this in evaluating the further polarisation by the specimen. The phenomenon is exactly the same as that which polarises light reflected from shiny surfaces and makes Polaroid glasses useful for cutting out glare.

Like the arc length factor, the polarisation factor leads to a relative weakening of rays reflected at medium angles around $2\theta = 90°$.

(v) *The Lorentz factor, L*

In the powder diffraction method it is assumed that particles are equally likely to lie in every orientation. In spite of this, the probability that a particle is appropriately oriented to give a reflection is not independent of the Bragg angle of the reflection concerned. This can be regarded as due to the fact that, for any real crystal, reflection is still possible for small finite mis-settings of the crystal (although the effect is independent of how small this tolerance may be). It is not possible to demonstrate clearly how this comes about or to derive a formula for this Lorentz factor, until after the concept of the reciprocal lattice has been introduced in Chapter 12, but if the orientations of the particles are truly random the intensities of the reflections are modified by a factor

$$L = \frac{1}{2 \sin \theta}.$$

Since this is a function of θ (not 2θ like the two previous angular factors) it diminishes all the way from the lowest to the highest angles.

(vi) *The absorption factor, A*

X-ray reflections are produced at all points within the specimen and the X-rays are attenuated by absorption as they travel through the specimen by a factor $\exp(-\mu x)$, where μ is the linear absorption coefficient of the specimen and x is the distance travelled in the specimen. The effect has to be summed numerically for all possible sites of reflection throughout the volume of the specimen, and therefore depends on the shape as well as the composition of the specimen. It is a function of θ, and although it cannot in general be expressed mathematically, tables are available for simple shapes such as cylinders and spheres. The value of μ has to be calculated for any particular specimen composition from tabulated values of atomic absorption coefficients and from the density.

There is an important exception to the statement that the absorption correction cannot in general be expressed mathematically. In the powder diffractometer method a flat specimen always makes an angle with the incident beam equal to the Bragg angle, and in these circumstances the path length traversed in reaching and leaving any point in the specimen, (AP + PB), is equal to $2t/\sin\theta$ (Fig. 10.7(a)). If the specimen is thick enough, rays reflected near the back of the specimen will be almost totally absorbed and the specimen can be regarded as infinitely thick. The contributions from all depths within the specimen can then be summed by means of an infinite integral

$$\int_0^\infty \exp(-2\mu t/\sin\theta)\,dt$$

$$= \sin\theta/2\mu.$$

Since the area of the specimen that is irradiated varies inversely with $\sin\theta$ because of the changing obliquity of the specimen to the beam (Fig. 10.7(b)), the angular factor cancels and the absorption factor can be regarded as a constant $1/\mu$, at all angles.

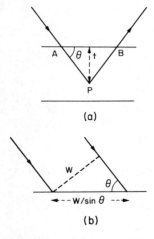

FIG. 10.7. Geometry of reflection from the specimen in a powder diffractometer: (a) the ray reflected at a depth t travels a distance $AP + PB = 2t/\sin\theta$ in the specimen, with corresponding absorption; (b) the beam of cross-section W irradiates an area $W/\sin\theta$ of the specimen.

In summary then, the intensity of a reflection hkl is proportional to

$$MILPA\ F_{hkl}^2$$

assuming that the temperature factor has been incorporated in the structure factor.

Problems

Metallic copper has a cubic unit cell containing four copper atoms at coordinates $0,0,0$; $0,\frac{1}{2},\frac{1}{2}$; $\frac{1}{2},0,\frac{1}{2}$; and $\frac{1}{2},\frac{1}{2},0$. The spacings of its 111 planes and 331 planes are 2.088Å and 0.829Å, respectively. The atomic scattering factor of copper is 22.1 at $\sin\theta/\lambda = 0.240$ (i.e. $1/2d_{111}$) and 11.5 at $\sin\theta/\lambda = 0.603$ ($1/2d_{331}$).

1. Calculate the structure factors F_{111} and F_{331}.
2. Calculate the multiplicity, arc length, polarisation and Lorentz factors appropriate to each of these reflections for (a) CuKα radiation and (b) MoKα radiation.
3. Hence find the ratio of the intensity of the 111 reflection to that of the 331 reflection as recorded on a powder diffractometer using (a) CuKα radiation and (b) MoKα radiation, assuming that the effect of the temperature factor can be neglected.

CHAPTER 11

The Fourteen Bravais Lattices

IN CHAPTER 3 we developed a classification of lattices, and the shapes of their unit cells, on the basis of their symmetry. To do this we started with the most general, triclinic, lattice based on a unit cell having all three edges a, b, c unequal and all three angles α, β, γ unequal, and we considered the effect of inserting various elements of symmetry into such a lattice. As a result we found seven kinds of lattice which we were able to take as the basis for classifying crystals into the seven crystal systems, and the discussion was sufficient for this purpose. It was not complete, however, because we did not examine all the orientations in which it is possible to introduce symmetry into lattices; when this is done it is found that there exist seven further kinds of lattice, making fourteen in all. A knowledge of all of these is necessary in interpreting X-ray diffraction patterns, and we therefore return to the subject at this point. The fourteen possible kinds of three-dimensional lattice are known collectively by the name of their discoverer, Bravais, who worked them out in 1848, long before they could be used in interpreting X-ray diffraction.

Monoclinic system

In developing the monoclinic lattice from the triclinic in Chapter 3 we inserted a mirror plane perpendicular to one set of parallel edges of the cell, those of length b (Fig. 3.5(b)). An alternative way of introducing the same degree of symmetry would be to put a mirror plane through one edge (say c) and bisecting the angle between the other two (Fig. 11.1(a)). The restrictions which this imposes on the shape of the new cell are to make $a = b$ and $\alpha = \beta$, but it does not require restriction of any angle to 90°.* Such a cell has the full symmetry characteristic of the monoclinic system, there being a 2-fold axis through the mid-points of the two opposite c-edges that do not lie in the mirror plane.

The reason why we did not need to take account of a lattice of this nature in discussing the external symmetry of monoclinic crystals is that it can be described in terms of a larger unit cell of conventional monoclinic shape ($a' \neq b' \neq c'$, $\alpha' = \gamma' = 90° \neq \beta'$) but with a lattice point at the centre of the C-faces as well as at the cell corners (Fig. 11.1(b)). Because of the right angles introduced between the two pairs of axes by this choice of cell it is very much more convenient to use and is almost always adopted. It is described as a C-centred unit cell, as distinct from a primitive cell which has lattice points only at its corners. It could

---

* The point made on p. 27 is to be borne in mind throughout this chapter. Symmetry is primary, and any equalities among the edges and angles of the cell are secondary. They are only significant if they are demanded by the symmetry.

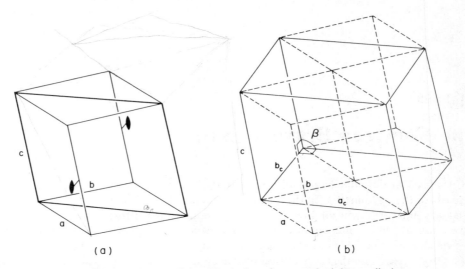

FIG. 11.1. (a) Unit cell with monoclinic symmetry but of non-standard shape; cell edges ———, mirror planes ———. (b) Lattice of unit cells as in (a), broken lines, described in terms of a cell of standard monoclinic shape centred on the C-face, full lines. Subscript c denotes edges of the centred cell.

equally well be described as A-centred depending on the choice of which cell edge is labelled a and which is labelled c, but the convention is to label the axes so as to make it C-centred unless there is some special reason to the contrary.

Because the new kind of lattice that we have just developed *can be* described as having a centred cell of ordinary monoclinic shape it is customary to describe it as a centred (C) monoclinic lattice, and the kind of monoclinic lattice discussed in Chapter 3 is described as a primitive (P) monoclinic lattice. This nomenclature is convenient and has become traditional, but it is strictly illogical. Both kinds of lattice possess primitive cells, and both are equally good monoclinic lattices in their own right. It would be equally possible to choose a centred cell in the "primitive monoclinic lattice" but it would have no particularly useful characteristics so we do not usually do so. The subject is further confused in most crystallography textbooks which develop the centred lattices by the curious consideration of "adding" lattice points to primitive cells. This gives unit cells a primacy over lattices that they do not possess; as we have already seen in Chapter 2 the lattice is the primary thing, and there is an infinity of different unit cells in terms of which any lattice can be described, the choice between them being purely one of human convenience.

There is indeed another unit cell that we can choose in a "centred monoclinic lattice" which also has the conventional monoclinic shape; it has the same b- and c-edges as the C-centred cell but its a-edge is the diagonal of the ac face of that cell (Fig. 11.2). It has a lattice point at its body centre instead of on a face, and is described as an I-centred cell (I for Inside). The C-centred and I-centred cells both have the same volume, which is twice that of the primitive cell of the centred lattice. There is always one lattice point per primitive cell, which may be regarded as "containing" $\frac{1}{8}$ share of the lattice point at each of its eight corners, since eight cells meet there. Similarly an I-cell obviously contains two

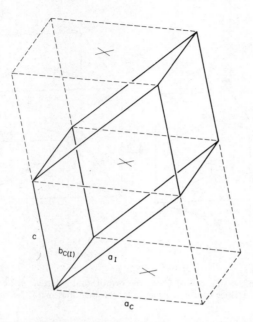

FIG. 11.2. Alternative description of the lattice of Fig. 11.1 in terms of a body-centred cell of standard monoclinic shape.

lattice points, a total of $8 \times \frac{1}{8}$ at its corners and one inside; and a C-cell also contains two, since the two face-centring points are each shared between two cells.

Orthorhombic system

In the orthorhombic system there are three new kinds of lattice with different orientations of the symmetry elements; they have non-rectangular cells, but may all be described in terms of centred rectangular cells of various kinds with the usual orthorhombic shape, $a \neq b \neq c$, $\alpha = \beta = \gamma = 90°$. To develop these in a logical way we start from the centred monoclinic lattice described in terms of its primitive cell, and insert the extra symmetry in three different orientations.

1. An additional mirror plane is inserted perpendicular to the c-edge in Fig. 11.1(a) thereby requiring $\alpha = \beta = 90°$. Thus the cell shape becomes $a = b \neq c$, $\alpha = \beta = 90° \neq \gamma$ and a third mirror plane arises parallel to c and perpendicular to the original one (Fig. 11.3(a)).

By a direct analogy with Fig. 11.1(b), a lattice of such cells can be described by a C-centred cell (Fig. 11.3(b)) in which all the edges are perpendicular to one another. This lattice is therefore described as C-centred orthorhombic.

The same process of insertion of an extra mirror plane can also be described in terms of its insertion perpendicular to the c-edge of the C-centred monoclinic cell of Fig. 11.1(b) thereby leading directly to a C-centred cell of conventional orthorhombic shape. Since in the orthorhombic system there is no strong preference as to which axis is which, the

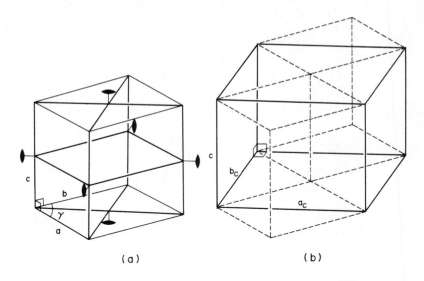

Fig. 11.3. (a) Primitive unit cell of C-centred orthorhombic lattice; cell edges ———, mirror planes ———. (b) Lattice of unit cells as in (a), broken lines, with C-centred cell, full lines.

C-centred lattice could equally, by an alternative labelling of axes, be described as A-centred or B-centred.

2. Although $\alpha = \beta = 90°$ is a necessary corollary of giving additional mirror symmetry to the primitive cell of the centred monoclinic lattice, there are two other ways of inserting this symmetry into the *lattice as a whole*. The first is to put the mirror plane perpendicular to the c-edge through the origin, O, and the opposite corner of the cell, N (Fig. 11.4(a)), thereby making the cell diagonal ON perpendicular to c. This mirror reflects A to A' and B to B' in the same cell, but it reflects C to the next lattice point below O and M to the next lattice point above N*. It automatically brings in a third mirror plane on ABB'A' which reflects O to N, but reflects C and M to vertices of adjacent cells. The lattice therefore has full orthorhombic symmetry, but the cell is constrained to have a shape given by

$$a = b, \ c = -2a \cos \alpha, \ \alpha = \beta \neq 90° \neq \gamma.$$

The lattice can, however, also be described in terms of an I-centred cell of conventional orthorhombic shape (Fig. 11.4(b)). In this system the C-centred and I-centred lattices are fundamentally different. The I-centred orthorhombic lattice can, of course, also be derived directly from the I-centred version of the centred monoclinic cell (Fig. 11.2) by insertion of a mirror plane perpendicular to its c-edge in the usual way; such a mirror is then in a different orientation in the lattice from that inserted in deriving the orthorhombic C-centred cell from the monoclinic C-centred cell.

---

* It must be remembered that a lattice is an array of points, not any particular set of lines joining those points together. The symmetry of the lattice is therefore the symmetry of the array of points not that of the lines, and there is, for example, no "cell edge" along AN corresponding to the reflection of A'N.

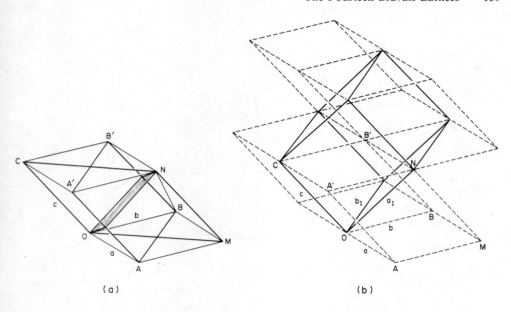

FIG. 11.4. (a) Primitive unit cell of I-centred orthorhombic lattice; cell edges ———, mirror planes ———; (b) Lattice of unit cells as in (a), broken lines, with I-centred cell, full lines.

3. The third possibility (Fig. 11.5(a)) is to insert mirror symmetry on the plane $11\bar{1}$ of the primitive cell of the centred monoclinic lattice (OA'B' of the Figure). This reflects C to N, and A, B and M to the next lattice points to the left. It constrains the diagonal ON to be equal to the C-edge and the next point above N to be vertically over O. The third mirror plane is therefore on the top face of the cell, and the shape of the cell is given by

$$a = b, \quad c = \frac{-a\cos^2\gamma/2}{\cos\alpha}, \quad \alpha = \beta \neq 90° \neq \gamma.$$

This lattice can also be described in terms of a centred cell of conventional orthorhombic shape but having lattice points at the centres of all six faces (Fig. 11.5(b)). It is called a face-centred (F-centred) cell, contains 4 lattice points ($8 \times \frac{1}{8} + 6 \times \frac{1}{2}$), and is four times the volume of the primitive cell. The F-centred orthorhombic cell can be related back to the centred monoclinic lattice, because that can also be described by an F-centred cell of conventional shape, though this is of course twice as large as the C- and I-centred cells and has no compensating advantage of simplicity, so it is not used. However, if a mirror plane is inserted perpendicular to the c-edge of such an F-centred monoclinic cell it will convert it to a conventional orthorhombic cell with F-centring.

There are thus four different kinds of lattice all having orthorhombic symmetry and called P-, C-, I- and F-lattices because their smallest unit cell having the conventional orthorhombic shape possesses centrings of these kinds.

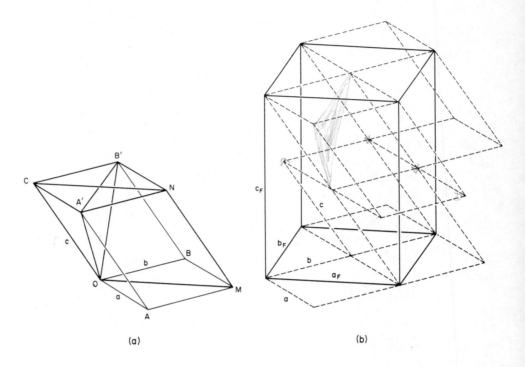

FIG. 11.5. (a) Primitive unit cell of F-centred orthorhombic lattice; cell edges ———, mirror planes ———. (b) Lattice of unit cells as in (a), broken lines, with F-centred cell, full lines.

Tetragonal system

Fortunately there are no additional complications arising in systems of higher symmetry than the orthorhombic; indeed there is no other system that has as many as four kinds of lattice. We can most simply derive the lattices in the tetragonal system in the same way as we did for the primitive lattice in Chapter 3 by putting a 4-fold rotation axis parallel to the c-edge of the centred orthorhombic cells. If this process is applied to the I-centred orthorhombic cell we obtain an I-centred tetragonal cell as in Fig. 11.6. However, if we do the same thing to a C-centred orthorhombic lattice the resulting lattice can be described by means of a smaller primitive cell of equally conventional tetragonal shape (Fig. 11.7); the lattice produced is therefore a P-lattice. Similarly, insertion of a 4-fold axis parallel to the c-edge in an F-centred orthorhombic lattice results in one that can be described by a smaller I-centred cell of conventional tetragonal shape (Fig. 11.8); this lattice is therefore an I-lattice, and it must be concluded that there are only two types of tetragonal lattice, P and I.

The shape of the primitive cell of the centred tetragonal lattice and the orientation of the symmetry elements with respect to it are noted in Table 11.1.

The Fourteen Bravais Lattices 161

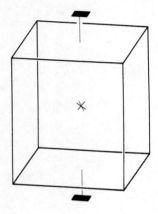

FIG. 11.6. I-centred tetragonal cell.

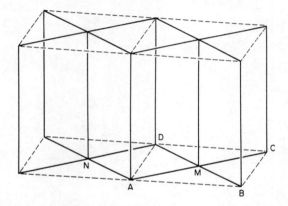

FIG. 11.7. A lattice of C-centred cells with tetragonal symmetry (broken lines) possesses a smaller primitive unit cell (full lines) which is also of standard tetragonal shape.

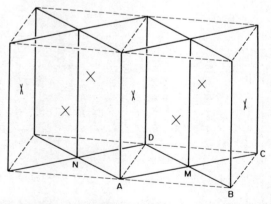

FIG. 11.8. A lattice of F-centred cells with tetragonal symmetry (broken lines) possesses a smaller I-centred unit cell (full lines) which is also of standard tetragonal shape.

TABLE 11.1. *Cells of the centred Bravais lattices*

System	Orientation of key symmetry elements in primitive unit cell	Constraints on primitive cell	Centring in smallest cell of standard form	Vector relations of centred cell relative to primitive cell	
Monoclinic	m bisects γ	$a = b \neq c$ $\alpha = \beta \neq 90° \neq \gamma$	C(A) or I	$\mathbf{a}_C = \mathbf{a} + \mathbf{b}$ $\mathbf{a}_I = \mathbf{a} + \mathbf{b} + \mathbf{c}$	$\mathbf{b}_C = \mathbf{b} - \mathbf{a}$ $\mathbf{b}_I = \mathbf{b} - \mathbf{a}$	$\mathbf{c}_C = \mathbf{c}$ $\mathbf{c}_I = \mathbf{c}$
Orthorhombic	m bisects γ m on plane 110 m on plane 002	$a = b \neq c$ $\alpha = \beta = 90° \neq \gamma$	C(A, B)	$\mathbf{a}_C = \mathbf{a} + \mathbf{b}$	$\mathbf{b}_C = \mathbf{b} - \mathbf{a}$	$\mathbf{c}_C = \mathbf{c}$
	m bisects γ *m on plane 110 *m on plane 11$\bar{2}$	$a = b, c = -2a\cos\alpha$ $\alpha = \beta \neq 90° \neq \gamma$	I	$\mathbf{a}_I = \mathbf{a} + \mathbf{b} + \mathbf{c}$	$\mathbf{b}_I = \mathbf{b} - \mathbf{a}$	$\mathbf{c}_I = \mathbf{c}$
	m bisects γ *m on plane 001 *m on plane 11$\bar{1}$	$a = b, c = -a\cos^2\frac{1}{2}\gamma/\cos\alpha$ $\alpha = \beta \neq 90° \neq \gamma$	F	$\mathbf{a}_F = \mathbf{a} + \mathbf{b}$	$\mathbf{b}_F = \mathbf{b} - \mathbf{a}$	$\mathbf{c}_F = \mathbf{a} + \mathbf{b} + 2\mathbf{c}$
Tetragonal	*4-axis $\parallel \mathbf{c}$ m bisects γ *m on planes 110, 11$\bar{2}$, 100, 010	$a = b, c = -2a\cos\alpha$ $\alpha = \beta = \sin^{-1}(\sqrt{2}\sin\gamma/2)$	I	$\mathbf{a}_I = \mathbf{a} + \mathbf{b} + \mathbf{c}$	$\mathbf{b}_I = \mathbf{b} - \mathbf{a}$	$\mathbf{c}_I = \mathbf{c}$
Cubic	*$\bar{3}$-axes along $a, b,$ [101], [011] *4-axes along c, [1$\bar{1}$0], [111]	$a = b, c = 2a/\sqrt{3}$ $\alpha = \beta = \cos^{-1}(-1/\sqrt{3})$, $\gamma = 2\sin^{-1}(1/\sqrt{3})$	I	$\mathbf{a}_I = \mathbf{a} + \mathbf{b} + \mathbf{c}$	$\mathbf{b}_I = \mathbf{b} - \mathbf{a}$	$\mathbf{c}_I = \mathbf{c}$
	*$\bar{3}$-axes along [132], [312], [$\bar{1}$12] *4-axes along [$\bar{1}$12], [1$\bar{1}$0], [110]	$a = b = c$ $\alpha = \beta = 120°$ $\gamma = 90°$	F	$\mathbf{a}_F = \mathbf{a} + \mathbf{b}$	$\mathbf{b}_F = \mathbf{b} - \mathbf{a}$	$\mathbf{c}_F = \mathbf{a} + \mathbf{b} + 2\mathbf{c}$

* denotes symmetry elements that are present in the lattice though not in an isolated unit cell. Throughout the table axes are chosen to make $\alpha \geq 90°$.

Cubic system

To derive the cubic Bravais lattices we proceed as in Chapter 3 to insert 3-fold axes along the diagonals of the centred orthorhombic cells. Clearly this requires all three faces meeting at a vertex to be identical, so the C-centred cell cannot be treated in this way, but P-, I- F-cubic cells can be derived. The shapes of the primitive cells in the I- and F-centred cubic lattices, and the orientation of the symmetry elements with respect to them are noted in Table 11.1.

Hexagonal and trigonal systems

In these systems no new types of lattice can be derived other than the hexagonal and rhombohedral lattices discussed in Chapter 3.

Summary of the Bravais lattices

The fourteen Bravais lattices are thus distributed very unevenly over the seven crystal systems as follows:

Triclinic	1
Monoclinic	2
Orthorhombic	4
Tetragonal	2
Cubic	3
Hexagonal Trigonal	2

It is to be noted that the division of a system into different Bravais lattices has no connection with its division into different crystal classes. A crystal in any class in a given system may have any of the Bravais lattices that belong to that system, and vice versa.

The relationship between the cell dimensions of a centred cell and of the primitive cell of the centred lattice which it describes is most simply expressed in terms of vectors, and it is given in this form in the last column of Table 11.1.

The effect of non-primitive cells in X-ray crystallography

In Chapter 2 we saw that the unit cell that we deduced from the morphology of a crystal might well differ in shape in various ways from a true unit cell of the lattice; for example, one of its edges, say c, might be twice the true value relative to the others, which is equivalent to taking a doubled unit cell, and this would not lead to any inconsistency in indexing the crystal faces. However, when we determine unit cell dimensions from X-ray diffraction measurements in the ways to be described in the following chapters, such ambiguities do not arise and we can always obtain a true *unit* cell of the lattice. It has therefore often happened that discrepancies have been found between axial ratios deduced from crystal morphology and unit cell dimensions deduced from X-ray measurements.

164   Crystallography for Earth Science Students

Thus the morphological axial ratios of an orthorhombic crystal might be $0.813:1:0.992$, and the X-ray unit cell might be found to be

$$a = 11.97 \text{ Å}, \qquad b = 14.72 \text{ Å}, \qquad c = 7.30 \text{ Å}.$$

Since a multiple of a unit cell is a cell of the lattice whereas a sub-multiple of it is not, the only way of interpreting the morphological axial ratios in terms of cell dimensions would be a cell with

$$a = 11.97 \text{ Å}, \qquad b = 14.72 \text{ Å}, \qquad c = 14.60 \text{ Å}.$$

In such a cell, for every atom at a position with fractional coordinates x, y, z, there would be a second similar atom at $x, y, z+\frac{1}{2}$, and therefore the structure factor F_{hkl} would be given by

$$F_{hkl} = \sum_i f_i \cos 2\pi(hx_i + ky_i + lz_i) + \sum_i f_i \cos 2\pi(hx_i + ky_i + lz_i + (\tfrac{1}{2}l))$$

where the summation is over the atoms in the true unit cell only.
Since

$$\cos(\theta + 2\pi) = \cos\theta$$

and

$$\cos(\theta + \pi) = -\cos\theta$$

the terms in the two sums will all cancel out if l is odd and reinforce one another if l is even. Thus if we adopted the morphological cell we would find a total absence of reflections with l odd because we would have indexed them on the basis of a non-primitive cell with a doubled c-edge.

A very similar situation arises in the case of centred Bravais lattices. A crystal having a centred Bravais lattice will give X-ray reflections corresponding to all its lattice planes, and if we were to index these on the basis of its primitive cell (of the form shown in Table 11.1) they would be found to have all possible indices in the usual way. But if, purely for our own convenience, we adopt what is really an artificially enlarged "unit" cell in order to make it a conventional shape, then we shall find that reflections with certain kinds of indices will be systematically missing.

The categories of missing reflections can be found in exactly the same way as above. In a C-centred cell, for every atom at x, y, z there will be another exactly similar atom at $x+\frac{1}{2}, y+\frac{1}{2}, z$ and the structure factor equation will be

$$F_{hkl} = \sum_i f_i \cos 2\pi(hx_i + ky_i + lz_i) + \sum_i f_i \cos 2\pi(hx_i + ky_i + lz_i + \tfrac{1}{2}(h+k))$$

where the summation is over the atoms in a primitive cell only. Hence the structure factor will be zero for $h+k$ odd, and reflections will only appear if $h+k$ is even.

Similarly in a body-centred cell we shall have a second atom at $x+\frac{1}{2}, y+\frac{1}{2}, z+\frac{1}{2}$, and

$$F_{hkl} = \sum_i f_i \cos 2\pi(hx_i + ky_i + lz_i) + \sum_i f_i \cos 2\pi(hx_i + ky_i + lz_i + \tfrac{1}{2}(h+k+l)).$$

Thus the structure factor is zero for $h+k+l$ odd and reflections only appear for $h+k+l$ even.

In a cell centred on all its faces there are corresponding atoms at x, y, z, at $x+\frac{1}{2}, y+\frac{1}{2}, z$, at $x+\frac{1}{2}, y, z+\frac{1}{2}$, and at $x, y+\frac{1}{2}, z+\frac{1}{2}$. The structure factor expression can therefore be divided into four summations

$$F_{hkl} = \sum_i f_i \cos 2\pi(hx_i + ky_i + lz_i) + \sum_i f_i \cos 2\pi(hx_i + ky_i + lz_i + \tfrac{1}{2}(h+k))$$
$$+ \sum_i f_i \cos 2\pi(hx_i + ky_i + lz_i + \tfrac{1}{2}(h+l)) \sum_i f_i \cos 2\pi(hx_i + ky_i + lz_i + \tfrac{1}{2}(k+l)).$$

As before, the first pair cancels if $h + k$ is odd, and a simple trigonometrical transformation shows that the second pair then also cancels. The symmetry of the expression clearly means that if we pair the first and third or first and fourth summations the expression will vanish if $h + l$ is odd or if $k + l$ is odd. Thus there will be an X-ray reflection only if $h + k$, $k + l$ and $l + h$ are simultaneously even. This can be expressed alternatively as the condition that h, k and l are *either* all even *or* all odd.

Rhombohedral and hexagonal cells in the trigonal system

It was pointed out in Chapter 3 that crystals in the trigonal system can always be described and indexed in terms either of a rhombohedral unit cell, or of a hexagonal unit cell whose contents have lower symmetry than corresponds to the hexagonal system. In some trigonal crystals the lattice is such that the rhombohedral cell is primitive and in order to define a hexagonal cell we have to take a non-primitive one, of three times the volume, having two lattice points at one-third and two-thirds of the way along a body diagonal (Fig. 11.9). Such a lattice is denoted as R. In other trigonal crystals the lattice is such that the hexagonal cell is primitive and in order to define a rhombohedral cell we have to take a non-primitive one, of three times the volume, having two lattice points at one-third and two-thirds of the way along its 3-fold axis (Fig. 11.10). Such a lattice is described as P because it is the same as a hexagonal P-lattice. Thus although there is no new kind of Bravais lattice in the trigonal system, it does contain two kinds of Bravais lattice, and they can be distinguished by the fact that the presence of additional lattice points within the non-primitive cells imposes restrictions on the indices of X-ray reflections for exactly the same reason as applies to the centred lattices of the other systems. The effect is that reflections are observed only when the indices satisfy the following conditions:

Lattice	Indexed on rhombohedral cell	Indexed on hexagonal cell
Rhombohedral (R)	no restriction	$-h + k + l = 3n$*
Hexagonal (P)	$p + q + r = 3n$	no restriction

* If the positive directions of the hexagonal x- and y- axes are reversed relative to the rhombohedral axes this condition becomes $h - k + l = 3n$, so that absence of reflections when either one or the other of these two conditions is not satisfied indicates an R-lattice.

Discussion of the relationship between the parameter c/a of a hexagonal type cell in the trigonal system and the parameter of the corresponding rhombohedral-type cell was postponed in Chapter 6 because it is complicated by the existence of these two kinds of lattice in the trigonal system. In an R-lattice the parameters of the rhombohedral and hexagonal cells are related as:

$$a_R^2 = \frac{a_H^2}{3} + \frac{c_H^2}{9} \quad \text{and} \quad \sin\frac{\alpha}{2} = \frac{3}{2\sqrt{\{3 + (c_H/a_H)^2\}}}.$$

In a P-lattice the corresponding relationships are

$$a_R^2 = a_H^2 + c_H^2 \quad \text{and} \quad \sin\frac{\alpha}{2} = \frac{3}{2\sqrt{\{3 + 3(c_H/a_H)^2\}}}.$$

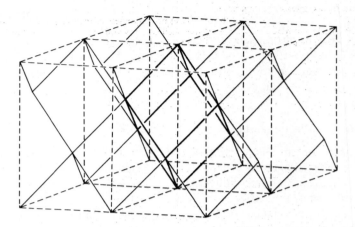

FIG. 11.9. Primitive rhombohedral lattice (full lines, one cell in bold lines) with alternative description in terms of a non-primitive hexagonal lattice (broken lines).

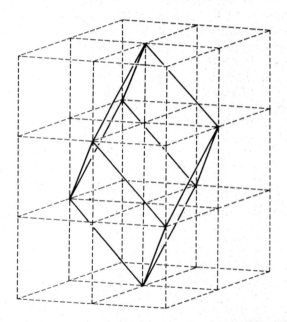

FIG. 11.10. Primitive hexagonal lattice (broken lines) with alternative description in terms of a non-primitive rhombohedral cell (full lines).

Problems

The tetragonal crystal described in Chapter 6, problem 6, was investigated by X-ray diffraction. With the form A assumed to be {100} it was found to have a unit cell having $a = 8.02$ Å, $c = 10.64$ Å. On the basis of this unit cell its X-ray reflections could be indexed as:

200, 400, 600, 220, 440, 660, 420, 620, 640, 111, 331, 551, 311, 511, 531, 202, 402, 602, 222, 422, 442, 622, 113, 333, 313, 513.

1. What Bravais lattice does this suggest?
2. What conclusions may be drawn about the relative appropriateness of the two possible indexings of form A?
3. What are the a- and c-dimensions of the smallest cell of conventional tetragonal shape?
4. In terms of this cell, what are the indices of the forms A, B, C, D?

CHAPTER 12

Interpretation of Powder Photographs

THE interpretation of any crystal diffraction pattern involves, as a first step, determining the indices of the lattice planes that can be regarded as reflecting each of the various diffracted rays. This process is termed indexing the diffraction pattern. In order to be able to do this we need to know the relationship between the interplanar spacing, d, of a set of planes and its indices, hkl. We shall derive this for orthogonal axes.

Figure 12.1 shows the first plane of a set hkl intercepting $a/h, b/k, c/l$ off the axes. ON, the perpendicular from the origin, is of length d. By considering the right-angled triangles OAN, OBN, OCN we have

$$\cos A\hat{O}N = \frac{dh}{a}, \quad \cos B\hat{O}N = \frac{dk}{b}, \quad \text{and} \quad \cos C\hat{O}N = \frac{dl}{c}.$$

But these are the direction cosines of the line ON and the sum of their squares is therefore unity.
Thus
$$\frac{d^2h^2}{a^2}+\frac{d^2k^2}{b^2}+\frac{d^2l^2}{c^2}=1,$$

so that
$$\frac{1}{d^2}=\frac{h^2}{a^2}+\frac{k^2}{b^2}+\frac{l^2}{c^2}. \tag{12.1}$$

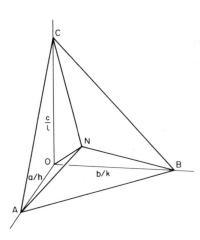

FIG. 12.1. ABC is the first plane from the origin of a set hkl. ON, the perpendicular from the origin on ABC, is of length d.

Powder patterns of cubic substances

In the case of a cubic crystal $a = b = c$, and eqn. (12.1) simplifies to

$$\frac{1}{d^2} = \frac{h^2 + k^2 + l^2}{a^2}. \tag{12.2}$$

If we express the Bragg Equation in the form

$$\sin\theta = \frac{\lambda}{2d},$$

then we can square it and substitute from eqn. (12.2) to give

$$\sin^2\theta = \frac{\lambda^2}{4a^2}(h^2 + k^2 + l^2). \tag{12.3}$$

The quantity $h^2 + k^2 + l^2$ is often called N, and it can take nearly all integral values, as can be seen from the following tabulation:

hkl	$N = h^2 + k^2 + l^2$	hkl	$N = h^2 + k^2 + l^2$
100	1	221 or 300	9
110	2	310	10
111	3	311	11
200	4	222	12
210	5	320	13
211	6	321	14
—	—	—	—
220	8	400	16

Note that there is no set of three integers whose squares sum to 7 or 15; in fact there is a series of such "forbidden" numbers whose next few members are 23, 28, 31, 39. . . . They can all be generated from the expression

$$4^p(8n+7)$$

where p and n are any positive integers or zero.

The above relationship makes it very easy to index a powder photograph or diffractometer trace of a cubic substance. The θ-values of the lines are measured, and converted to values of $\sin^2\theta$. In principle these values should then be found to be in arithmetic progression with common difference $\lambda^2/4a^2$ except where the sequence is interrupted by a forbidden value of N, and there the difference between successive terms should be twice as great. The values of N should therefore be determinable simply by counting along the sequence, and then interpreted into terms of the corresponding hkl values. The value of the cell edge a could then be found from the value of $\sin\theta$ for any line for which N was known.

In practice the procedure is not quite so straightforward as this for two reasons, absent lines and errors in measuring θ. Some reflections may have a zero, or very small, structure factor, so that the corresponding lines are unobserved and lead to gaps in the regular progression of $\sin^2\theta$ values. The possibility of such gaps must always be taken into account. Thus the smallest intervals in $\sin^2\theta$ between successive lines should be sought first, as providing a probable value of $\lambda^2/4a^2$, and larger intervals (spanning gaps in the sequence due to absent lines) should be interpretable as multiples of this value. The presence of errors in θ means that a certain amount of judgement has to be exercised in making the interpretation. In the example shown in Table 12.1 a preliminary estimate of $\lambda^2/4a^2 = 0.030$ is obtained from the first three differences, and this is divided into each of the first five values of $\sin^2 \theta$ to give approximate values of N. These run consistently above

TABLE 12.1 *Indexing of a primitive cubic powder photograph taken with $FeK\alpha$ radiation:*
λ mean = 1.937 Å; λ_{α_1} = 1.9359 Å; λ_{α_2} = 1.9399 Å

θ	$\sin^2 \theta$	Differences	approx. N	True N	hkl
18.5	0.101		3.3; 3.2	3	111
		0.030*			
21.2	0.131		4.4; 4.1	4	200
		0.032*			
23.8	0.163		5.4; 5.1	5	210
		0.028*			
25.9	0.191		6.3; 6.0	6	211
		0.064			
30.3	0.255		8.5; 8.0	8	220
		0.099			
36.5	0.354		11.1	11	311
		0.025			
38.0	0.379		11.8	12	222
		0.038			
40.3	0.417		13.0	13	320
		0.031			
42.0	0.448		14.0	14	321
		0.156			
51.0	0.604		18.9	19	331
		0.039			
53.3	0.643		20.1	20	420
		0.030			
55.1	0.673		21.0	21	421
57.0	0.703		21.96 ⎱	22	332
57.1	0.705		22.03 ⎰		
61.0	0.765		23.91 ⎱	24	422
61.2	0.768		24.00 ⎰		
68.1	0.861		26.91 ⎱	27	511/333
68.4	0.864		27.00 ⎰		
74.2	0.926		28.94 ⎱	29	520/432
74.5	0.929		29.03 ⎰		
78.2	0.958		29.94 ⎱	30	521
78.6	0.961		30.03 ⎰		

The first three differences (marked *) suggest a first approximation of 0.030 for $\lambda^2/4a^2$. This is divided into the first five values of $\sin^2 \theta$ and gives the first approximate values of N. A second approximation is chosen (0.032) to give more nearly integral values. The final value of $\lambda^2/4a^2$ is taken as 0.958/30 from the $K\alpha_1$ line with $N = 30$, giving $a = 5.417$ Å.

integral values, so a second estimate of $\lambda^2/4a^2 = 0.032$ is derived on the assumption that the fourth and fifth reflections have $N = 6$ and 8. Approximate values of N calculated from this agree well with permissible integers. The high-angle lines above $N = 21$ are resolved into their $K\alpha_1$ and $K\alpha_2$ components, and were not included when calculating the column of differences. However, their indexing follows easily, and a final value of a is calculated from the $K\alpha_1$ component of the line with $N = 30$ using the accurate value of the wavelength of $FeK\alpha_1$ radiation.

Special cases arise if the specimen has a centred Bravais lattice because then there is a systematic pattern of absent reflections corresponding to $h + k + l$ odd for an I-lattice or to mixed odd and even indices for an F-lattice. It can be readily verified that if $h + k + l$ is to be even then $h^2 + k^2 + l^2$ ($= N$) must also be even, so that a body-centred cubic substance gives a pattern of lines based on even values of N only. The absences for an F-centred cubic substance due to mixed odd and even indices are more extensive; the pattern contains only lines for which $N = 4n$ or $N = 4(2n+1) - 1$. Thus the three kinds of cubic Bravais lattice lead to the following sequences of N values up to $N = 32$:

P	1	2	3	4	5	6	X	8	9	10	11	12	13	14	X	16
I		2		4		6		8		10		12		14		16
F			3	4				8			11	12				16
P	17	18	19	20	21	22	X	24	25	26	27	X	29	30	31	32
I		18		20		22		24		26		X		30		32
F			19	20				24			27	X				32

where X denotes the position of the "forbidden" values of N.

The F-centred pattern is very distinctive, with its regular succession of three differently sized intervals between the $\sin^2 \theta$ values corresponding to increments in N of 3, 1 and 4. The close pairs with unit increment in N provide a clear indication of the value of $\lambda^2/4a^2$. The I-centred pattern is a little more confusing, because there does not exist any pair of lines with a unit difference in N, and at first sight the regular sequence of $\sin^2 \theta$ values may be mistaken for that which would be given by a P-lattice having a value of $\lambda^2/4a^2$ twice as big. The distinction can be made on the basis of the "forbidden" values of N: the seventh and fifteenth members of the series from a P-lattice would be forbidden, whereas the seventh member of the series from an I-lattice has $N = 14$ and the fifteenth has $N = 30$, so that both are permitted. On the other hand, the fourteenth member, which is permitted for a P-lattice, is forbidden for an I-lattice because it has $N = 28$.

A value of the unit cell dimension, a, can be obtained from the $\sin \theta$ value of any line for which N has been determined using eqn. (12.3). However, in order to obtain an accurate value of a from a powder photograph, lines at the highest possible values of θ should be used. At high values of θ approaching 90° various systematic sources of error (such as displacement of the lines by absorption effects) tend to zero, and the effect on $\sin \theta$ of random errors in measurement of line position also tends to zero since

$$\frac{d \sin \theta}{d\theta} = \cos \theta.$$

For the most accurate work, values of a calculated from a series of very high-angle lines on a powder photograph may be plotted against a suitable function* of θ and extrapolated to $\theta = 90°$. This procedure is only justified if the measurement of the film is done to the highest degree of accuracy by the method described on p. 132, and if the lines on the film are sharp and well defined.

In powder diffractometry it is not possible to record reflections at θ values near to $90°$, but errors are generally eliminated by use of an internal standard, mixed with the specimen and giving lines at accurately known positions. For example, Table 12.2 shows the θ-values for the first eleven lines of a pattern as recorded on a diffractometer from a mixture of a cubic mineral with powdered silicon as an internal standard. The d-spacings of silicon are known, and so the true θ-values of its lines can be calculated (θ_{calc}). The errors, $\Delta\theta(Si)$ of the θ-values of the silicon lines, can be plotted on a graph against θ_{obs}, and interpolated values of $\Delta\theta$ for the unknown lines can then be subtracted from θ_{obs} to give θ_{corr}. The values of $\sin^2\theta$ derived from each line are then of good accuracy, and lead to unambiguous $N(h^2 + k^2 + l^2)$ values. The mineral evidently has a face-centred cubic lattice, and reasonably consistent values of the unit cell edge a are obtained from each line.

Although the indexing of a cubic powder pattern from a table of the values of $\sin^2\theta$ for the lines is always possible, it is often unnecessary, because the appearance of the patterns

TABLE 12.2 *Correction of power diffractometer peaks from a cubic mineral using an internal standard (silicon)*

θ_{obs}	θ_{calc}(Si)	$\Delta\theta$(Si)	Correction	θ_{corr}	$\sin^2\theta$	$\Delta(\sin^2\theta)$	N	a(Å)
9.83			0.36	9.47	0.0271		3	8.12
14.66	14.24	0.42				0.0458		
16.09			0.43	15.66	0.0729		8	8.079
						0.0273		
18.93			0.48	18.45	0.1002		11	8.080
						0.0089		
19.78			0.49	19.29	0.1091		12	8.086
						0.0364		
22.95			0.53	22.42	0.1455		16	8.085
24.23	23.68	0.55						
*28.59	(28.09)		0.61	(27.98)	(0.2201)		(24)	(8.05)
						0.1003		
30.36			0.64	29.72	0.2458		27	8.081
						0.0451		
33.31			0.67	32.64	0.2909		32	8.086
35.31	34.61	0.77						

* This line is an incompletely resolved superposition of a silicon line and a line of the pattern. It is therefore not used in determining either $\Delta\theta$ or a.

Inevitable random errors in reading the line positions account for the scatter in the results for a. These have the greatest effect at low angles, so the first line is ignored. The mean value of a from the other good lines is 8.083 ± 0.001 Å. A more precise value could only be obtained from lines at higher angles.

* The most linear, and therefore the most accurate, extrapolation is considered to be obtained by plotting the values of a against the values of the function $\left(\dfrac{\cos^2\theta}{\sin\theta} + \dfrac{\cos^2\theta}{\theta}\right)$ for the lines from which they are derived.

is so characteristic that they can be indexed by inspection. Figure 12.2 shows a graph of $\sin^2\theta$ against θ on which intervals of θ are marked corresponding to equal intervals of $\sin^2\theta$. Although the intervals in θ are not equal, they are very regular, and in the middle of the range they are very nearly equal to one another. As a result of this it is possible to recognise by eye the gaps between lines on a powder photograph or diffractometer trace as corresponding to particular increments in the corresponding N values by comparison with the smallest gap that occurs. Examples of powder photographs of materials with P-, I- and F-lattices are shown in Fig. 12.3(a–c) to make this clear. In Fig. 12.3(a) there is a sequence of lines at regular intervals, from which the seventh member is missing; this is sufficient to characterise it as having a P-lattice. In Fig. 12.3(b) there is again a regular sequence of lines, and although it is difficult to allocate N-values immediately because the first one or two members of the series are missing, it is clear that the seventh member is present. This shows that the material does not have a P-lattice and so must have an I-lattice. The first gap in the sequence must therefore be at the position of the fourteenth member of the sequence with $N = 28$. This establishes the N-values for all the lines. Figure 12.3(c) shows the sequence of lines characteristic of an F-lattice and so may be indexed directly.

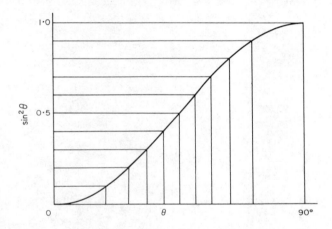

FIG. 12.2. Intervals of θ corresponding to equal intervals of $\sin^2\theta$.

Powder patterns of non-cubic substances

The interpretation of cubic powder patterns is simple because such patterns can be described in terms of only one variable parameter. Thus the sequence of lines in terms of their indices is invariant (100, 110, 111, 200, etc.), even though some of them may be absent. All powder photographs are therefore like Fig. 12.3(a–c) except for variations in relative intensities of the lines (due to the dependence of the structure factor on the atomic arrangement) and a scale factor ($\lambda^2/4a^2$) that brings the lines closer together or opens them out in terms of a $\sin^2\theta$ scale. However, as soon as we go to an even slightly less symmetrical crystal system this simplicity is completely lost.

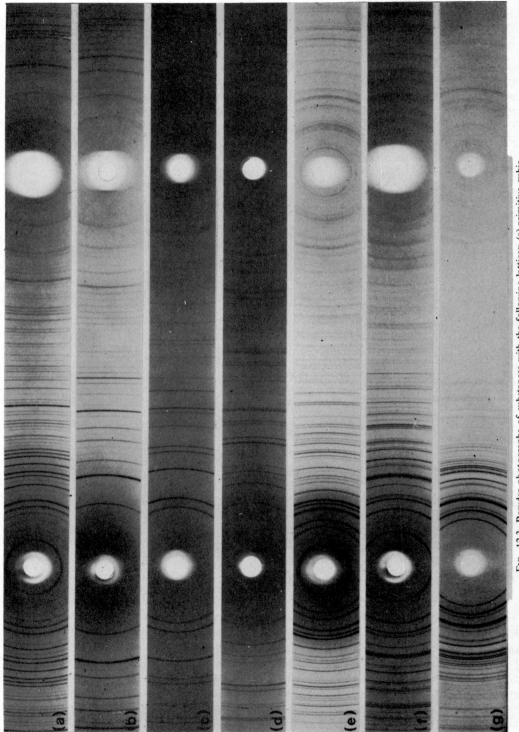

Fig. 12.3. Powder photographs of substances with the following lattices: (a) primitive cubic, (b) body-centred cubic, (c) face-centred cubic, (d) tetragonal, (e) hexagonal, (f) trigonal, (g) orthorhombic.

For the tetragonal system we may put $a = b$ in eqn. (12.1), and hence we can derive

$$\sin^2 \theta = \frac{\lambda^2}{4a^2}(h^2 + k^2) + \frac{\lambda^2 l^2}{4c^2} \tag{12.4}$$

as the analogue of eqn. (12.3). The loss of equality between a and c complicates the result in two ways. Firstly there are more reflections, because $\sin^2 \theta$ is different when l is interchanged with h or k (e.g. the reflection 101 is in a different position from 110). Secondly, and much more importantly, the order of the reflections is no longer constant: for example, 200 may be at any distance before or after 111 depending on the relative values of a and c. The extra complexity is shown in Fig. 12.3(d), but it is impossible for this to typify tetragonal powder photographs in the way that Fig. 12.3(a–c) typify cubic powder photographs.

In a table of values of $\sin^2 \theta$ corresponding to the lines of a tetragonal powder photograph there exist regular increments of $\lambda^2/4a^2$ within sets of lines having constant l^2 and varying $h^2 + k^2$, and there exist other regular increments of $\lambda^2/4c^2$ within sets of lines having constant $h^2 + k^2$ and varying l^2, but there is no obvious way of distinguishing these sets, and there are many irregular increments between pairs of lines where both $h^2 + k^2$ and l^2 vary. Such a photograph can be indexed by a laborious search among the differences in $\sin^2 \theta$ between all possible pairs of lines (not just adjacent pairs) for frequently occurring values, and multiples of them. Such a search may be programmed on a computer, and trial and error graphical methods also exist for fitting the patterns to appropriate values of a and c. These methods will not be discussed here.

If sufficient prior information is available about the approximate unit cell dimensions of the specimen, either from the application of the above methods to other similar (but not identical) specimens of the same material, or by the single crystal methods described in Chapter 13, then expected values of $\sin^2 \theta$ for the various reflections can be calculated. These can then be compared with the observed values and the equivalence may be sufficiently good to permit the indices of the observed lines to be identified. If this can be done, then new refined values of a and c can be derived from the observed values of $\sin^2 \theta$. The simplest way is to use a high-angle line with $l = 0$ to find a and a high-angle line with $h^2 + k^2 = 0$ to find c, but more sophisticated methods will use a range of reflections and a least-squares method to find the best values of a and c to fit them all simultaneously.

Similar remarks apply to other lower symmetry systems. In the hexagonal system (and the trigonal system indexed on a hexagonal unit cell) $\sin^2 \theta$ is given by

$$\sin^2 \theta = \frac{\lambda^2}{3a^2}(h^2 + hk + k^2) + \frac{\lambda^2 l^2}{4c^2}. \tag{12.5}$$

There are the same number of variables (a and c) as in the tetragonal system and the problem of indexing is of equal complexity. There are still more lines in the pattern, however, because the term hk in the expression leads to the reflections hkl and $h\bar{k}l$ occurring at different angles. Examples are shown in Fig. 12.3(e) and (f). In the orthorhombic system we have to substitute eqn. (12.1) itself into the square of the Bragg Equation. There are three unknowns a, b and c, and the complexity is such that indexing is only practicable by computer. An example is shown in Fig. 12.3 (g). In the monoclinic and triclinic systems the formulae become still more complex, with four and six variables respectively.

Ambiguities in indexing powder reflections

In indexing a cubic powder pattern we identify the reflections in terms of $N (= h^2 + k^2 + l^2)$, but these values of N cannot always be uniquely identified with a particular set of indices hkl. We have already seen that $N = 9$ may correspond to the indices 221 or 300 so that a line with $N = 9$ will consist of a superposition of reflections from these two sets of planes. They are unrelated to one another structurally, so their intensities may be quite different, but we cannot record them separately or even find out whether one or other of them is absent. Such ambiguities become increasingly frequent as N increases and above $N = 50$ they predominate over unique correspondences of N with specific sets of indices. In some cases there may be more than two possible sets of indices for the same value of N; for example, $N = 41$ corresponds to 621, 540 and 443.

Another possible kind of ambiguity has been concealed so far because we have not considered permutations of the hkl values and reversals of their signs. In the holosymmetric class of the cubic system all such permutations and reversals of sign lead to sets of planes that are symmetrically equivalent to one another, and the intensities of the corresponding reflections are therefore identical. Thus no ambiguity arises, and the intensity corresponding to any one set of such planes is the observed intensity divided by the multiplicity factor discussed in Chapter 10. However, in the lower symmetry classes of the cubic system these sets of planes are not all symmetrically related and in some cases they divide up into two or four non-equivalent sub-sets corresponding to the non-equivalent forms of lower multiplicity discussed in Chapter 7. However, the sub-sets always possess identical d-spacings, and the corresponding non-equivalent reflections are superimposed and cannot be isolated from one another by powder photography or powder diffractometry.

In the lower symmetry systems the number of such systematic superpositions of different reflections is reduced, but because of the increased number and irregular disposition of the reflections accidental superpositions occur, and these become very common if the unit cell is large.

Both of these kinds of ambiguity can be resolved by single crystal diffraction which is discussed in Chapter 13.

Problems

1. A cubic mineral gives a powder photograph with lines at the following measured values of θ with CrKα radiation ($\lambda = 2.291$ Å):

    14.1°, 17.4°, 24.9°, 27.2°, 29.2°, 33.1°, 34.7°, 38.4°, 41.8°, 43.5°.

    Deduce the Bravais lattice and the size of unit cell.
2. Another cubic mineral gives diffractometer peaks at the following measured θ values with CoKα radiation ($\lambda = 1.790$ Å):

    10.6°, 17.6°, 20.7°, 21.7°, 25.2°, 31.5°, 33.6°, 37.1°, 39.1°, 42.4°, 44.3°.

    Deduce the Bravais lattice and the size of unit cell.
3. Sylvite (potassium chloride) gives powder lines with CuKα radiation ($\lambda = 1.542$ Å) at θ values of 14.2°, 20.3°, 25.1°, 29.3°, 33.2°, 36.9°, 43.9°, 47.3°, 50.8°, 54.4°, 58.1°, 62.1°, 66.5°, 78.6°.

    What Bravais lattice and cell size does this suggest? Halite (sodium chloride) has a unit cell with $a = 5.64$ Å. It has a face-centred cubic lattice, and if the origin of the cell is taken at a sodium ion there are Na$^+$ ions at the vertices and centres of the faces in contact with Cl⁻ ions at the mid-points of the edges of the cell. The structure of sylvite would be expected on chemical grounds to be similar, and K⁺ ions have a radius that is about 0.3 Å larger than that of Na⁺ ions. If this expectation is correct, what are the true indices of the sylvite reflections? Suggest a reason for the misleading nature of the powder pattern.

CHAPTER 13

X-ray Diffraction by Single Crystals

Difficulties associated with powder photographs

POWDER diffraction photographs constitute a very convenient method of investigating cubic substances. As we have seen in Chapter 12, a cubic substance with any one of the three cubic Bravais lattices gives a pattern of lines on a powder photograph which is invariant except for a scale factor that is controlled by the edge length of the unit cell. The indices of the sets of planes which give rise to the lines can usually be recognised immediately by inspection of the pattern, and even if this is made difficult by an unexpected absence or weakness of particular lines the problem is readily solved by analysing a table of $\sin^2 \theta$ values of the lines. Once the indices of the lines have been found the size of the unit cell can be obtained very simply, and from the intensities of the lines it is then possible to find the arrangement of the atoms within the unit of pattern in the unit cell by the methods to be described in Chapter 15.

However, when a substance belongs to a crystal system of lower symmetry than the cubic its powder pattern is much more complicated. This does not matter when we use the powder photograph for the purpose of identification as discussed in Chapter 9; indeed, it is then a positive advantage because the complexity makes the pattern more individually characteristic of the substance to be identified. But the complexity makes it very difficult to index the lines unless the cell dimensions are known approximately, and therefore to use such a pattern to determine either cell dimensions or atomic arrangements. Even in the cubic system there are problems due to superposition of non-equivalent reflections if the crystal class is not holosymmetric, as discussed in Chapter 12.

The reason for the complexity can be seen from the expression for the interplanar spacings of an orthorhombic crystal (eqn. (12.1))

$$\frac{1}{d_{hkl}^2} = \frac{h^2}{a^2} + \frac{k^2}{b^2} + \frac{l^2}{c^2}.$$

When this is substituted into the squared form of the Bragg equation we obtain:

$$\sin^2 \theta_{hkl} = \frac{\lambda^2}{4}\left(\frac{h^2}{a^2} + \frac{k^2}{b^2} + \frac{l^2}{c^2}\right) \tag{13.1}$$

instead of the corresponding expression for the cubic system (eqn. (12.3))

$$\sin^2 \theta_{hkl} = \frac{\lambda^2}{4a^2}(h^2 + k^2 + l^2).$$

The order of the lines derived from eqn. (12.3) is invariable; for example, the reflection 200 is always at a higher angle than 110 because $2^2 > 1^2 + 1^2$. But there is no such invariance about the order of reflections from an orthorhombic material that are governed by eqn. (13.1), because (to consider the same pair of reflections)

$$\frac{2^2}{a^2} \gtrless \frac{1^2}{a^2} + \frac{1^2}{b^2}$$

depending on the relative values of a and b. It is therefore impossible to index powder photographs of an orthorhombic mineral by inspection. It is even impossible to tell by inspection of the photograph that the mineral is orthorhombic. Either of these processes would require very extensive computational analysis of a table of the $\sin^2 \theta$ values, and even with an appropriate computer program the success of a solution is very dependent on the accuracy of the measurements. For measurements of cell dimensions, and for determination of structures, powder photographs of non-cubic substances are therefore very disadvantageous, unless the indices of the lines have already been identified in some other way.

Rotation photographs

A great simplification in interpretation can be achieved by using a single crystal instead of a powder. The crystal is set up with a particular crystallographic axis (say the z-axis) vertical and a horizontal beam of monochromatic X-rays is directed at it, as in Fig. 13.1. In general there may be no set of crystallographic planes at the appropriate Bragg angle to the beam to reflect the X-rays. However, if the crystal is rotated about the z-axis, then crystallographic planes parallel to the z-axis will take up sequentially all possible angles to the beam, and at appropriate moments during the rotation they will give reflections, all of which will lie in the plane through the crystal perpendicular to the z-axis, the equatorial plane of the experiment. These crystallographic planes, since they are parallel to z, necessarily have indices of the type $hk0$; any planes with a non-zero l index which are able to give Bragg reflections will necessarily reflect the beam out of the equatorial plane. Thus if we record the reflections on a photographic film, placed cylindrically round the crystal,

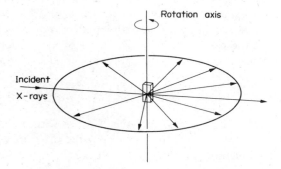

FIG. 13.1. A crystal rotated about its vertical z-axis produces all its $hk0$ reflections in a horizontal plane.

there will be a row of spots on the equatorial line which will contain all the $hk0$ reflections and no others. An example of such a single crystal *rotation photograph* is shown in Fig. 13.2.

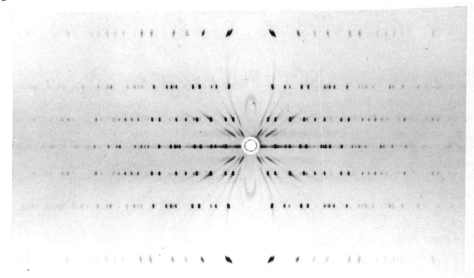

FIG. 13.2. A single crystal rotation photograph (of rutile).

This technique not only sorts out the $hk0$ reflections from all the others (or, of course, the $0kl$ or $h0l$ reflections if we had chosen the x- or y-axis as the rotation axis perpendicular to the beam), it also provides us with a direct means of measuring the unit cell dimension that is parallel to the rotation axis. In order to appreciate this we must forget, temporarily, the whole concept of Bragg reflections from lattice planes, and consider the much more basic idea of diffraction by a single row of equally spaced atoms as shown in Fig. 13.3. If the crystal has been set with the z-axis vertical, then every atom in the crystal will be part of just such a vertical row of equally spaced atoms, a distance c apart. One such row is shown in Fig. 13.3, together with a ray scattered from it at an angular elevation v above the horizontal (i.e. lying anywhere on a cone of semi-vertical angle $90°-v$ with the row of atoms along its axis). Because of the difference in path length travelled by rays scattered from successive atoms along the row, such rays will destructively interfere with one another unless the path difference is a whole number of wavelengths, i.e. unless

$$\sin v = \frac{n\lambda}{c}. \tag{13.2}$$

This relationship defines a set of cones as shown in Fig. 13.4 with values of $n = 0, \pm 1, \pm 2$. The rays diffracted from every vertical row of atoms in the structure will suffer destructive interference except in directions that lie on these cones, and there can therefore be no diffracted rays from the crystal as a whole except in directions lying on them. Of course any diffracted X-rays must also satisfy the Bragg equation for some particular set of lattice planes, and for this reason we do not get reflections everywhere round the cones, but

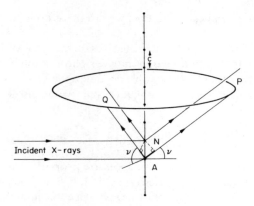

FIG. 13.3. Diffraction from a row of equally spaced atoms. The path difference (AN = $c \sin v$) for rays scattered by adjacent atoms is constant round the cone PQ, and for constructive interference must equal $n\lambda$.

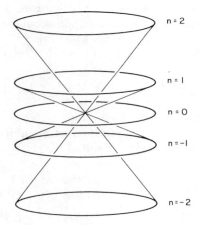

FIG. 13.4. A set of cones given by $\sin v = n\lambda/c$ with $n = 0, \pm 1, \pm 2$.

nevertheless all reflections must lie on them. The reflections from the crystal therefore give a series of spots on a cylindrical film which lie on horizontal lines when it is opened out. The horizontal lines of spots recorded on the film are called *layer lines*. The equatorial one, which we have already shown to contain the $hk0$ reflections, is called the *zero layer line*, and the others are called the first (upper or lower) layer line, the second (upper or lower) layer line, etc. If the radius of the camera is R and the height of the nth layer line above the zero layer line is Y on the film, then

$$\tan v = \frac{Y}{R}. \qquad (13.3)$$

Thus we can find v and hence from eqn. (13.2) we can find c. By taking further rotation photographs about the x- and y-axes in turn we can therefore find directly all three edges of the unit cell.

With the $hk0$ reflections segregated by themselves on the zero layer line it becomes much easier to index them than it would be on a powder photograph. On Fig. 13.2, which is a rotation photograph of rutile, it is possible to index the zero layer line by inspection, because rutile is tetragonal. For a tetragonal crystal $a = b$, and so from eqn. (13.1) with $l = 0$

$$\sin^2 \theta_{hk0} = \frac{\lambda^2}{4a^2}(h^2 + k^2). \tag{13.4}$$

Thus the values of $\sin^2 \theta$ for the successive spots are proportional to all the integers that can be expressed as the sum of two squares, i.e. 1, 2, 4, 5, 8, 9, 10, 13. ... The process is exactly like that of indexing a powder photograph of a cubic mineral, except that there are many more "forbidden" values of $h^2 + k^2$ than there are of $h^2 + k^2 + l^2$. One must of course also allow for the possibility that some reflections may be too weak to be seen on the photograph, and in fact the spots corresponding to $h^2 + k^2 = 1$ and 9 are missing on Fig. 13.2.

An alternative way of indexing the zero layer line is to do it graphically. If we have already measured a by taking a rotation photograph about the x-axis of the crystal, we can draw a net of squares, each of side λ/a, as shown in Fig. 13.5. Then the distance from the origin of the net (O) to the point (h, k) of the net (P) is given by

$$\text{OP} = \sqrt{\left\{\left(\frac{h\lambda}{a}\right)^2 + \left(\frac{k\lambda}{a}\right)^2\right\}}$$

and from eqn. (13.4) this is $2 \sin \theta$ for the reflection $hk0$. Thus if we measure $2 \sin \theta$ for each spot on the zero layer line of the photograph and set a pair of dividers to this value on the same scale as the drawing we can find the point of the net at that distance from the origin and read off its h and k values from the net.

A great virtue of this graphical method is that it can be applied to crystals of any symmetry. Even for monoclinic crystals the value of $2 \sin \theta$ is given by

$$2 \sin \theta = \sqrt{\left\{\left(\frac{h\lambda}{d_{100}}\right)^2 + \left(\frac{k\lambda}{d_{010}}\right)^2 + \left(\frac{l\lambda}{d_{001}}\right)^2 - 2\left(\frac{h\lambda}{d_{100}}\right)\left(\frac{l\lambda}{d_{001}}\right)\cos\beta\right\}}$$

and with $l = 0$ this is equal to the distance from the origin to the point (h, k) of a net drawn as in Fig. 13.6(a) with cells measuring $\lambda/d_{100}(= \lambda/a \sin \beta)$ and $\lambda/d_{010}(= \lambda/b)$. Even on the zero layer line of a rotation photograph taken about the y-axis, the only change that has to be made is shown in Fig. 13.6(b); the axes are drawn at an angle $(180°-\beta)$, but the cells still measure $\lambda/d_{100}(= \lambda/a \sin \beta)$ and $\lambda/d_{001}(= \lambda/c \sin \beta)$.* For triclinic crystals the only further complication lies in rather complex formulae for the values of d_{100}, etc., and for the angles between the axes of such diagrams.

A diagram of the kind used in Figs. 13.5 and 13.6 is so useful in X-ray crystallography that it is dignified with a name, the *reciprocal lattice*. This name is based on the fact that it is a lattice whose unit translations bear a reciprocal relationship to the dimensions of the

---

* Only the positive quadrant is shown of Fig. 13.6(b), but in fact two quadrants would be required because the distance from the origin to the point $h0l$ is not equal to that to $h0\bar{l}$.

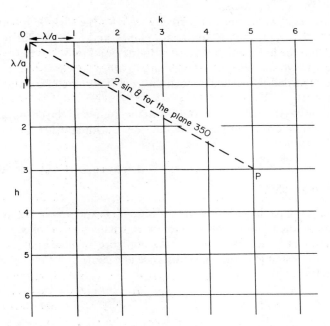

Fig. 13.5. Net for graphical indexing of the reflections on the zero layer line of rutile in terms of their values of $2\sin\theta$.

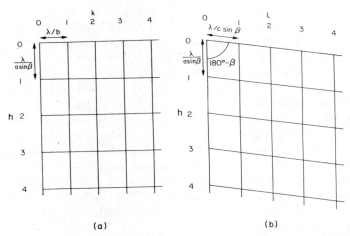

Fig. 13.6. Nets corresponding to Fig. 13.5 for indexing the reflections on the zero layer line of rotation photographs of a particular monoclinic crystal:
(a) for rotation about the z-axis;
(b) for rotation about the y-axis.

lattice of the crystal. The values of λ/d_{100}, λ/d_{010} and λ/d_{001} are conventionally and conveniently referred to as a^*, b^*, c^*, respectively, and the directions of the axes along which these distances are set off can be conveniently called x^*, y^*, z^*. The angles between these reciprocal axes always come out correctly, even for the triclinic system, if x^* is drawn perpendicular to the 100 planes, y^* to the 010 planes, and z^* to the 001 planes; and it can then be shown that the line joining the origin to the point (h, k, l) is always perpendicular to the hkl planes, as well as being equal in length to λ/d_{hkl}. The plane nets we have used up to now can be regarded as particular sections of a three-dimensional reciprocal lattice defined in this way.

The method used to index reflections on the non-zero layer lines of a rotation photograph is best deferred until after we have dealt with another kind of single crystal photograph, the oscillation photograph.

Oscillation photographs*

Although the principle of indexing the reflections on the zero layer line of a rotation photograph is very much simpler than that of indexing any powder photograph other than one of a cubic mineral, nevertheless it can in practice involve some ambiguities. For example, even with a simple tetragonal substance like rutile, it is totally impossible to tell whether a reflection with

$$h^2 + k^2 = 25$$

is the 340 or the 500 reflection, since

$$3^2 + 4^2 = 5^2 + 0^2.$$

In lower symmetry systems ambiguities are less absolute, but nevertheless there are often approximate equalities between the calculated values of d_{hk0} for different values of h and k. Also, if the cell is large, so that there are many reflections, these will quite frequently overlap and it becomes difficult to identify them individually for certain.

These problems can be solved by oscillating the crystal through a limited angle about its vertical axis during the photographic exposure instead of rotating it. If it is rotated, all possible planes parallel to the rotation axis will pass through their Bragg angle during the rotation and will therefore contribute a spot on the zero layer line. If the crystal is oscillated through a limited angle some planes will be brought into their reflecting position and others will not; then the crystal can be turned further and a separate photograph taken while the crystal oscillates through a further angular range, thereby recording reflections from other sets of planes. Provided that we can predict which planes will reflect within a particular range of angular positions of the crystal relative to the incident X-ray beam, we can then resolve all ambiguities. This prediction can be made by a simple graphical method with the help of the reciprocal lattice.

---

* The use of oscillation photographs was widely superseded by moving-film methods such as the Weissenberg goniometer and the precession camera many years ago, although there has been some increase in its use more recently. However, the interpretation of oscillation photographs provides a much better introduction to the reciprocal lattice and is directly relevant to the interpretation of textures and of electron diffraction patterns.

It is for this reason that it is treated here in detail. Those requiring information on moving-film methods are referred to the books of M. J. Buerger, *X-Ray Crystallography* (Wiley, 1942) and *The Precession Method in X-Ray Crystallography* (Wiley, 1964).

In Fig. 13.7 a crystal at C has its z-axis perpendicular to the page and a set of planes $hk0$ is in the Bragg reflecting position for an X-ray beam incident along AC. The reflected ray travels along CP. A circle of unit radius is drawn around C as centre, AC is produced to cut the circle at O, and OP is joined. A line CN parallel to the reflecting planes cuts OP at N. We then have:

$$\hat{OCN} = \hat{NCP} = \hat{OAP} = \theta \text{ (the Bragg angle)};$$

$$\hat{APO} = 90° \text{ (the angle in a semicircle)};$$

and since CN is parallel to AP (both make an angle θ with AO) OP is perpendicular to CN and therefore to the reflecting planes. Also from the right-angled triangle AOP, since we have made AO = 2,

$$OP/2 = \sin \theta,$$

or

$$OP = 2 \sin \theta = \lambda/d_{hk0}.$$

Thus OP is equal, in both length and orientation, to the line joining the origin of the reciprocal lattice to the point $(h, k, 0)$.

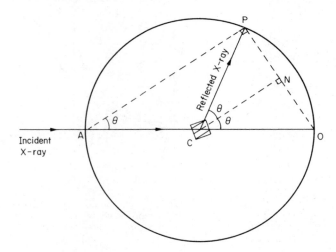

FIG. 13.7. A two-dimensional solution of the Bragg Equation for a reflection on the zero layer line. AO = 2; OP = λ/d and is perpendicular to the reflecting planes.

The consequence of the above analysis is that we have a graphical method of solving the Bragg equation. If we place the origin of the $hk0$ section of the reciprocal lattice at O, and orient it correctly in relation to the crystal, then there will be a Bragg reflection from the $hk0$ planes of the crystal if, and only if, the point $hk0$ of the reciprocal lattice lies on a unit circle through O whose diameter is on the line of the incident X-ray beam. Thus, during any given oscillation of the crystal at C, the planes that will reflect are those corresponding to the points of the reciprocal lattice that cross this circle during a corresponding oscillation of the reciprocal lattice about O.

The most practical way of applying this is to imagine the X-ray beam to oscillate with respect to the crystal instead of vice versa. Then we can keep the reciprocal lattice fixed while the beam, and the associated circle of unit radius, moves from AO to A'O in Fig. 13.8. During this oscillation the points of the reciprocal lattice that lie in the shaded area will have passed through the circle, and the corresponding crystal planes will have had the opportunity to reflect. Therefore we can index a reflection on the zero layer line of an oscillation photograph by seeking a point in the shaded area of the reciprocal lattice which lies at a distance from O equal to the value of $2 \sin \theta$ for the reflection. By suitably limiting the angle of oscillation we can therefore eliminate ambiguities of indexing.

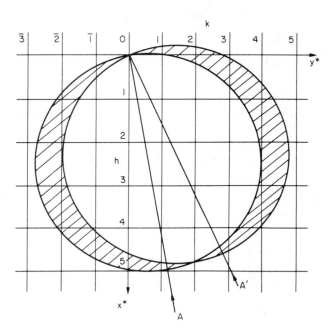

FIG. 13.8. x^*y^* section of a reciprocal lattice. The X-ray beam oscillates from AO to A'O and the $hk0$ planes that can reflect during the osicllation are represented by the points of the reciprocal lattice through which the circle of reflection passes, i.e. those in the shaded area.

Non-zero layer lines

Up to this point we have only considered the problem of indexing the zero layer line, and for this purpose we have used the section of the reciprocal lattice perpendicular to the rotation or oscillation axis of the crystal. However, the way in which we defined the reciprocal lattice on p. 184 was valid in three-dimensional terms, and Fig. 13.7 can be generalised into three dimensions in order to deal with any reflection, whether on the zero layer line or not. Figure 13.9 shows such a generalisation. The reflected ray is in a general direction CP, and instead of a circle of unit radius we have a sphere of unit radius centred at C. Exactly the same argument applied to the triangle AOP shows that OP = $2 \sin \theta$, that P is a point of the reciprocal lattice, and that a set of planes hkl in the crystal reflects if and

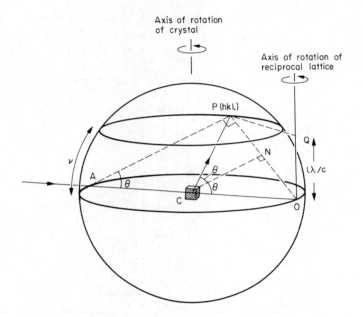

FIG. 13.9. Three-dimensional graphical solution of the Bragg Equation for any reflection. AO = 2. OP = λ/d and is perpendicular to the reflecting planes. QP = ξ, OQ = ζ, the coordinates of the reflection on the Bernal chart.

only if the point hkl of the reciprocal lattice lies on the sphere, which is therefore called the *sphere of reflection*.

All the points $hk1$ are at a perpendicular height λ/c above the plane of the $hk0$ points.*
Thus the $hk1$ points always intersect the sphere round a line of latitude at an elevation v above the horizontal given by

$$\sin v = \frac{\lambda}{c}.$$

Similarly the points hkl intersect the sphere where

$$\sin v = \frac{l\lambda}{c}.$$

This value of v is that of the lth layer line, so that it is now evident that the reflections on each successive layer line of a rotation or oscillation photograph about the z-axis have the value of the l index equal to the number of the layer line. If the rotation axis were x (or y) then the value of h (or k) would be given in the same way.

It would be difficult to index a reflection on a non-zero layer line by seeking a point of the reciprocal lattice at a distance from the origin equal to the $2 \sin \theta$ value of the reflection, because this would require a three-dimensional model of the reciprocal lattice. However,

---

* This is obvious if z is an orthogonal axis, but it can be shown to be true in general even for inclined axes when $c^* \neq \lambda/c$ (see Fig. 13.11).

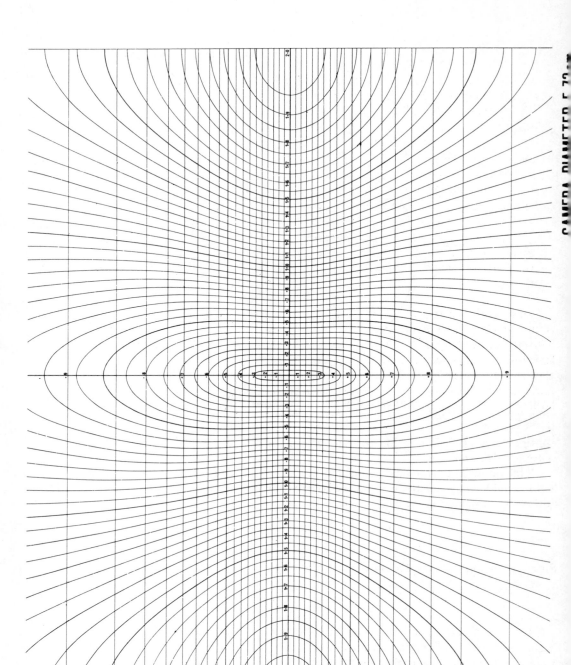

FIG. 13.10. The Bernal Chart constructed on the appropriate scale to measure the ξ, ζ reflections on a photograph taken on a cylindrical camera of diameter 57.3 mm.

we can do the operation in two dimensions on a drawing if we confine our attention to one plane of the reciprocal lattice at a time. The radial distance from the rotation axis to a point on the *l*th plane of the reciprocal lattice is given by

$$QP = \sqrt{\left(4\sin^2\theta - \frac{l^2\lambda^2}{c^2}\right)}$$

as is evident from Fig. 13.9. If we use this quantity instead of $2\sin\theta$ itself, then we can identify the indices of reflections on any layer line. This quantity is given the symbol ξ and it can be measured directly on a rotation or oscillation photograph by the use of a Bernal chart (Fig. 13.10) which is contoured in terms of ξ (the approximately vertical curves). The vertical scale is in terms of $l\lambda/c$ (or $h\lambda/a$ or $k\lambda/c$ as the case may be), and this coordinate is usually symbolised by ζ. Thus the ξ, ζ coordinates of any spot on a photograph can be read directly from the Bernal chart, and

$$\xi^2 + \zeta^2 = 4\sin^2\theta.$$

Indexing a non-zero layer line of a rotation photograph taken about an orthogonal axis is then an identical process to indexing a zero layer line. On a drawing of the reciprocal lattice section at height $l\lambda/c$ (which is, of course, identical with the net used for the zero layer line) a point is sought at a distance equal to the measured value of ξ from the rotation axis. If the rotation axis is not an orthogonal axis of the crystal then one has to take account of the fact that the rotation axis will not coincide with an axis of the reciprocal lattice. Since ξ is the radial distance from the rotation axis it has to be measured from the point where this cuts the particular level of the reciprocal lattice, and not from the origin of that level. The way of allowing for this is shown in Fig. 13.11 for a monoclinic crystal. The

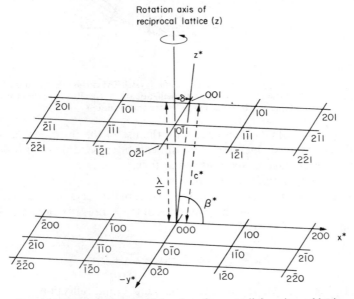

FIG. 13.11. Three-dimensional drawing of a monoclinic reciprocal lattice. On the *l*th layer the origin of the layer is displaced from the rotation axis by
$$\delta = lc^* \cos\beta^*.$$

190   Crystallography for Earth Science Students

displacement of the origin of the level from the rotation axis is rather more complicated in the case of a triclinic crystal.

In indexing non-zero layerlines of oscillation photographs the possible points of the reciprocal lattice are again confined between the two extreme positions in which the sphere of reflection intersects it at the beginning and end of oscillation. However, the sphere of reflection intersects a non-zero level of the reciprocal lattice at height ζ in a circle of radius $\sqrt{(1-\zeta^2)}$ (Fig. 13.9). The centre of this circle lies vertically above the centre of the sphere, at unit radial distance from the oscillation axis. Thus the circle no longer reaches the oscillation axis and the possible points are confined to the shaded region between two circles as shown in Fig. 13.12. It may be noted that no points at $\xi < 1 - \sqrt{(1-\zeta^2)}$ can pass through the sphere, and this corresponds to the way in which the lines of constant ξ on the Bernal chart curve round so that even those spots at the vertical centre-line of a photograph have non-zero values of ξ.

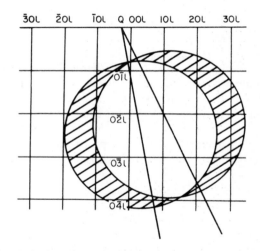

Fig. 13.12. Indexing the *l*th layer line of a monoclinic crystal. The circles are of radius $\sqrt{(1-\zeta^2)}$ and are centred at unit distance from the axis of oscillation (Q), vertically above the centre of the sphere of reflection in its extreme positions.

Thus single crystal methods, combined with the use of the reciprocal lattice concept, enable us to do two things which are often difficult or impossible by powder methods:
1. to determine each edge of the unit cell in turn;
2. to index all the reflections unambiguously.

This is the first step towards determining the structure of a crystal. If we are to find the atomic coordinates (x_i, y_i, z_i) from the intensities of the X-ray reflections via the structure factor equation

$$F_{hkl} = \sum_i f_i \cos 2\pi (hx_i + ky_i + lz_i)$$

we must clearly know the values of *hkl* for each individual reflection.

Another use of the reciprocal lattice is in a sense the converse of that which we have considered so far. For example, an acicular crystal may have ill-developed faces, and if we take an oscillation photograph about its long axis we shall not know the orientation of the crystal to the beam. However, if we know the unit cell dimensions we can draw the reciprocal lattice and index the spots on the photograph merely from their ξ values, marking the corresponding points on the reciprocal lattice, as in Fig. 13.13(a). The region of the reciprocal lattice occupied by the marked points then indicates the area through which the sphere of reflection has oscillated, and the orientation of the crystal to the beam can be deduced as in Fig. 13.13(b). An extension of this procedure is useful in interpreting X-ray diffraction from oriented aggregates of very small crystals. Such specimens are often of importance in mineralogy. Unlike powder specimens they do not have crystals disposed

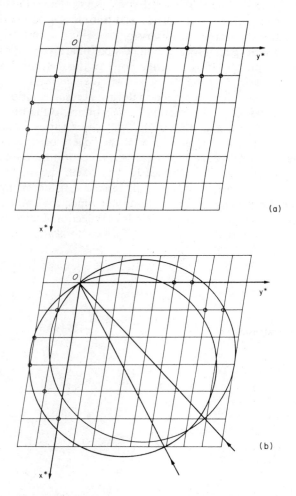

FIG. 13.13. An oscillation photograph of a crystal gives reflections on the zero layer line at $\xi = 1.00$, 1.10, 1.38 and 1.58 with deviations to the right of the incident beam and $\xi = 0.39$, 0.78, 1.05, 1.27 and 1.50 to the left. The reciprocal lattice is shown in (a), and the points at distances from O corresponding to these ξ values are marked in. The range of positions of the sphere of reflection, and therefore of the incident beam relative to the reciprocal lattice axes, can then be deduced to be as shown in (b).

with equal frequency in all possible orientations. For example, in a fibre texture the crystals may be in all possible orientations around one specific crystallographic axis, and the resulting X-ray diffraction pattern will then be identical with that obtained from a rotating single crystal, and can be interpreted in the same way by using the reciprocal lattice. Other intermediate situations between a perfect single crystal orientation and random disorientation are made much more tractable by treatment in terms of the reciprocal lattice. Provided that a few of the reflections from a partially oriented aggregate can be identified (from their $2 \sin \theta$ values), then it is possible to deduce the range of orientations of the component crystals that is required to bring the points of their reciprocal lattices corresponding to these reflections on to the sphere of reflection. At the same time the possibility of other ranges of orientation can be excluded because they would lead to reflections that are not observed.

The concept of the reciprocal lattice also makes possible an elucidation of the Lorentz factor discussed in Chapter 10. If we regard a point of the reciprocal lattice as having an infinitesimal (but non-zero) diameter Δ, then it will take an infinitesimal (but non-zero) time δt to pass through the sphere of reflection during a crystal rotation or oscillation. The intensity of any given reflection will then be proportional to the time δt for which that reflection is being produced, and the way in which this varies from one reflection to another can be evaluated from the geometry of the sphere of reflection.

To illustrate the principle we take the case of a reflection on the zero layer line of a rotation photograph. The time δt will clearly depend on the angle subtended by the reciprocal lattice point at the origin and this will decrease with increasing distance from the origin. However, δt also depends on the obliquity at which the point passes through the sphere, i.e. in this case the angle at which the circle of reflection intersects a circle centred at O and of radius ξ. As may be deduced from Fig. 13.14 the total angle of rotation $\delta \omega$, during which the crystal can reflect will be given by

$$\delta \omega = \frac{PP'}{2 \sin \theta}.$$

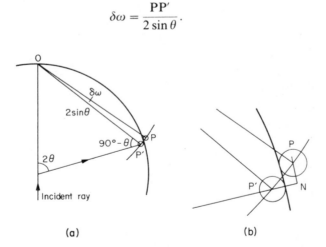

FIG. 13.14. Derivation of the Lorentz factor for the zero layer line of a rotation or oscillation photograph. (a) The point P of non-zero diameter Δ, passing through the circle of reflection. (b) Enlargement of region PP'.

But $P'N = \Delta$ and $P\hat{P}'N = \theta$
so that $PP' = \Delta/\cos\theta$, and

$$\delta\omega = \frac{\Delta}{2\sin\theta\cos\theta}.$$

Thus the Lorentz factor, which expresses the relative effect on the observed intensity arising from this cause, is $2\sin\theta\cos\theta$ ($=\sin 2\theta$) for reflections on the zero layer line of an oscillation or rotation photograph. In the case of a fibre texture, where there is no actual rotation, the same factor takes account of the fraction of the small crystals in the specimen that are in the appropriate orientation to reflect. In powder diffraction similar considerations apply but the situation is equivalent to the reciprocal lattice point approaching the sphere of reflection with equal probability for all orientations of the rotation axis. For every different geometrical arrangement of a diffraction experiment the appropriate Lorentz factor may be worked out from the same principles.

The reason why the reciprocal lattice is so valuable, in spite of its apparently abstract nature, is that it replaces the concept of the orientation in space of a set of planes by the much simpler concept of the position of a point. A further application in which the use of the reciprocal lattice is quite essential is the interpretation of electron diffraction in the electron microscope, which is discussed in Chapter 16.

Problems

A tetragonal crystal is mounted to rotate about its z-axis and a rotation photograph is taken with CuKα radiation ($\lambda = 1.542$ Å) on a camera of diameter 57.3 mm.

1. There are layer lines on the photograph at distances of 7.4 mm and 16.5 mm above and below the zero layer line. Calculate the c-dimension of the unit cell.
2. On the zero layer line there are spots at the following values of ξ as measured with a Bernal chart: 0.29, 0.41, 0.58, 0.65, 0.82, 0.87, 0.92, 1.04 and 1.15. Using the fact that on the zero layer line $\xi = 2\sin\theta$, suggest indices for these reflections and calculate the a-dimension of the corresponding unit cell.
3. On the first layer line there are spots at ξ-values of 0.20, 0.46, 0.61, 0.74, 0.84, 1.02 and 1.10 while on the second layer line there are spots at the same ξ-values as on the zero layer line. What modifications to the conclusions reached in question 2 are suggested by these reflections? Index all the reflections, deduce the Bravais lattice and calculate the true a-dimension of the unit cell.
4. A monoclinic mineral known to have a unit cell with $a = 6.95$ Å, $b = 4.72$ Å, $c = 8.17$ Å, $\beta = 99°$, forms a fibrous texture. A fragment is set with its fibre axis parallel to the axis of a cylindrical single-crystal camera, and a photograph taken with CuKα radiation ($\lambda = 1.542$ Å), specimen stationary. The photograph obtained looks like an oscillation photograph. There are layer lines at $\zeta = 0.325$ and $\zeta = 0.65$. On the right-hand side of the zero layer line there are reflections at ξ-values of 0.38, 0.57, 0.76, 0.83, 1.02, 1.12 and 1.30 and on the left-hand side at ξ-values of 0.41, 0.67, 0.94, 1.08 and 1.20. Which crystallographic axis is parallel to the fibre axis? Use the reciprocal lattice construction to determine the range of angles between the X-ray beam and one of the other crystallographic axes.

CHAPTER 14

Symmetry in Repeating Patterns

THE symmetry operations that were discussed in Chapter 3 were those applicable to finite objects like the external form of a crystal. When we consider repeating patterns (like crystal structures) that can be considered to extend indefinitely in all directions we have to take account of some additional kinds of symmetry operation, and it is necessary to examine the nature of these before considering how crystal structures can be determined.

The reason for the existence of these additional kinds of symmetry in repeating patterns is inherent in the definition that a symmetry operation moves or transforms an object in such a way that after transformation it coincides with itself. Imagine a crystal structure model of the conventional ball-and-spoke type which was a model of a whole crystal, with all the repetition that that implies, instead of a model of just one or two unit cells. Such models are usually built to a scale of 2.5 cm ≡ 1 Å; i.e. 2.5×10^8 times natural size. Thus a complete model of a 1-mm size crystal would extend for 250 km in all directions. If such a model were translated by one unit cell in any direction (perhaps 20 cm or so) it could for all practical purposes be said to coincide with itself: this would be true for every atom within it, and only contravened for those at the surface. Thus for infinite repeating patterns translations equal to multiples of, or vector additions of, the cell edges must be regarded as symmetry operations. Translations are therefore also permissible as components of more complex symmetry operations.

The simplest symmetry operation that has a translation component is the 2-fold screw axis, whose effect on a pattern that repeats in one direction only is shown in Fig. 14.1. The symmetry operation is rotation through 180° around the axis combined with translation of a half of the repeat distance of the pattern parallel to the axis. If the operation is

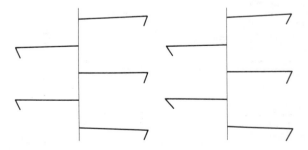

FIG. 14.1. The symmetry produced by a 2_1-screw axis. See p. 27 for instructions for viewing the stereoscopic drawings in Figs. 14.1–14.8.

performed twice the effect is equivalent to a simple translation by one period of the pattern. The symbol used for a 2-fold screw-axis is 2_1.

Higher-order screw axes are symbolised as n_m where the rotation is through $360/n°$ and the translation is through m/n of a period, where m is any positive integer less than n. Thus after n operations the total effect in every case is rotation through $360°$ (equivalent to no rotation) plus a simple translation by m periods of the lattice. The possibilities for crystallographic symmetry are restricted to $2_1, 3_1, 3_2, 4_1, 4_2, 4_3, 6_1, 6_2, 6_3, 6_4, 6_5$. Of these 3_1-, 4_1-, and 6_1-screw axes are very similar in nature to 2_1, as may be seen from Figs. 14.2-14.4 each repetition of the motif being at a different angle and a different level. 3_2-, 4_3- and

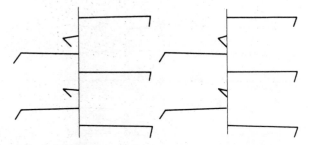

FIG. 14.2. The symmetry produced by (a) a 3_1-screw axis, and (c) a 6_1-screw axis.

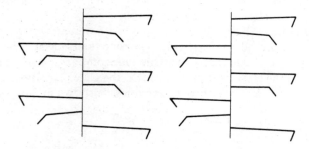

FIG. 14.3. The symmetry produced by a 4_1-screw axis.

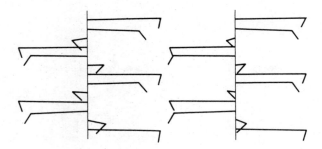

FIG. 14.4. The symmetry produced by a 6_1-screw axis.

196   Crystallography for Earth Science Students

6_5-screw axes differ only in the "hand" of the rotation, since translations by 2/3, 3/4 and 5/6 of a period upwards are equivalent to translations by 1/3, 1/4 and 1/6 of a period downwards, and are therefore not illustrated. A 4_2-screw axis introduces a different effect, however, as shown in Fig. 14.5. After two operations of the symmetry the motif is at level 0 but 180° from the initial position; after one and three operations of the symmetry the motif is at the level $\frac{1}{2}$, but respectively at orientations 90° and 270° from the initial position. Thus the pattern contains ordinary 2-fold symmetry as a component of the 4_2-symmetry. Similarly a 6_2-screw axis introduces ordinary 2-fold symmetry with pairs of motifs at levels 0, 1/3 and 2/3, and a 6_3-screw axis introduces ordinary 3-fold symmetry with three motifs at levels 0 and $\frac{1}{2}$. A 6_4-screw axis is the same as 6_2 but of opposite hand.

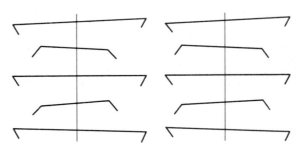

FIG. 14.5. The symmetry produced by a 4_2-screw axis incorporates 2-fold rotation symmetry.

A translational element may also be combined with mirror symmetry to give a glide plane of symmetry. The motif is reflected in a plane and then moved by half a period of the pattern in a direction parallel to that plane. This glide component may be parallel to either of those cell-edges of the crystal that are parallel to the glide plane, or to a diagonal of the cell face that is parallel to it. These three possibilities are shown in Figs. 14.6–14.8, and are

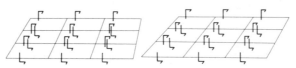

FIG. 14.6. The symmetry produced by an *a*-glide plane.

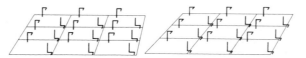

FIG. 14.7. The symmetry produced by a *b*-glide plane.

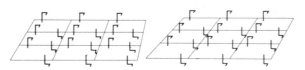

FIG. 14.8. The symmetry produced by a *n*-glide plane.

the only possibilities if the cell is primitive, since after two operations of the symmetry the movement of the motif must coincide with a simple lattice translation. These three types of glide plane are symbolised as a-, b- or c-glide planes if the translation is parallel to the x-, y- or z-axis (and equal to $\frac{1}{2}\mathbf{a}$, $\frac{1}{2}\mathbf{b}$, or $\frac{1}{2}\mathbf{c}$) and as an n-glide plane if it is parallel to a face-diagonal (and equal to $\frac{1}{2}(\mathbf{a}+\mathbf{b})$, $\frac{1}{2}(\mathbf{b}+\mathbf{c})$ or $\frac{1}{2}(\mathbf{c}+\mathbf{a})$). However, if the pattern is face-centred then it is also possible for the glide-component to be $\frac{1}{4}$ of the face diagonal, so that after two operations the motif is moved to the face-centring position. This is called a d-glide plane. It is really an artefact arising from the artificial (but convenient) use of centred instead of primitive cells.

If the pattern of the structure of a crystal contains a screw axis or a glide plane in a particular orientation then the morphology of the crystal will exhibit an ordinary axis of symmetry (of the same order) or a mirror plane, respectively, in the same orientation. Thus a mirror-plane in the morphology of a crystal may correspond *either* to a mirror plane *or* to some kind of glide plane in the structure, and an ordinary n-fold axis in the morphology may correspond *either* to an ordinary n-fold axis *or* to some kind of n-fold screw axis in the structure. Just as there is only a limited number of self-consistent ways of combining the morphological crystallographic symmetry elements with one another to give the (32) point-groups, so there is a limited number of self-consistent combinations of the structural symmetry elements to give what are called *space-groups*. However, because of the greater number of different kinds of symmetry elements in this case the number is much larger, namely 230.

Because, as we have seen, the lattice translations of a repeating pattern are part of its symmetry, the specification of a space-group requires first of all the specification of its Bravais lattice P, C, I or F. This symbol is then followed by a sequence of symbols for the axes and planes of symmetry in the structure, the order of which follows exactly the same conventions as in the symbol of a point-group (p. 84). For example, a particular tetragonal space-group is $P4_2/n\,2_1/m\,2/c$. This has a primitive lattice, with a 4_2-screw axis parallel to z and an n-glide plane perpendicular to it, a 2_1-screw axis parallel to x and a mirror plane

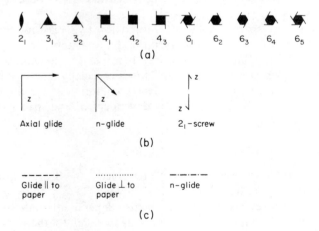

FIG. 14.9. The symbols used to denote (a) screw axes perpendicular to the paper, (b) glide planes (and 2_1-screw axis) parallel to the paper at height z above it, and (c) glide planes perpendicular to the paper.

perpendicular to it, and a 2-fold rotation axis in the direction at 45° to the x-axis and a c-glide plane perpendicular to it. It is very easy to see what morphology will be exhibited by a crystal that has a particular space-group simply by replacing all screw axes by corresponding ordinary axes of symmetry and all glide planes by mirrors. Thus a crystal with space-group $P4_2/n\ 2_1/m\ 2/c$ belongs to crystal class $4/m\ 2/m\ 2/m$. Just as point-group nomenclature is often abbreviated by omission of the symbols of those symmetry elements whose presence is implied by the others, so also is space-group nomenclature. Thus the above space-group can be written as $P4_2/n\ m\ c$ and the class as $4/m\ m\ m$.

For a complete list of the 230 space-groups the reader is referred to such books as *Chemical Crystallography* by C. W. Bunn (Oxford, 1945) or *X-ray Crystallography* by M. J. Buerger (Wiley, 1942). It suffices here to list all the space-groups belonging to the three monoclinic classes in order to indicate the sort of range of symbols that is introduced in rather simple cases in going from point-group to space-group.

	Point group		
Bravais lattice	2	m	2/m
P	P2 P2$_1$	P*m* P*c*	P2/m P2$_1$/m P2/c P2$_1$/c
C	C2	C*m* C*c*	C2/m C2/c

It is to be noted that not all permutations of the symmetry elements are mutually compatible, so that the list is in fact shorter than might be expected. Also, that it is conventional in the monoclinic system to list the symbols for an orientation of the crystal axes that puts the z-axis in the direction of any glide component; with a different orientation of axes one could equally well have P2/a, etc. Such an orientation would necessarily require a centred lattice to be in the A-centred orientation; e.g. C2/c would be converted to A2/a.

An understanding of space-groups is required for two reasons. First because the determination of the space-group of a crystal is the first step towards determining its structure on the atomic scale. And secondly because it is a necessary basis for an understanding of known crystal structures. This is especially true in mineralogy where substitution of one element by another is very common, and the limitations on such substitutions and their effects are often very dependent on the space-group.

Space-group determination

The first element of the space-group to be determined is the Bravais lattice, and this has been dealt with already in Chapter 11.

The detection of screw axes and glide planes depends on very similar principles. Suppose that a crystal contains a 2_1-screw axis parallel to the z-axis (though not necessarily *on* the z-axis). Then an atom at x, y, z will be moved by the operation of the screw to a related position at $x', y', z+\frac{1}{2}$. Thus every atom in the structure that has a coordinate z is paired with another similar atom at different x- and y-coordinates but at

$z+\tfrac{1}{2}$. Thus the structure factor for the hkl reflections will be

$$F_{hkl} = \sum_i f_i \{\cos 2\pi(hx_i + ky_i + lz_i) + \cos 2\pi(hx'_i + ky'_i + l(z_i + \tfrac{1}{2}))\}.$$

For reflections of type $00l$ this becomes

$$F_{00l} = \sum_i f_i \{\cos 2\pi lz_i + \cos 2\pi l(z_i + \tfrac{1}{2})\}.$$

$$= 2\sum_i f_i \cos 2\pi lz_i \text{ for } l \text{ even}$$

and zero for l odd.

It is to be noted that such cancellation of the terms when l is odd occurs only for reflections of type $00l$, with two zero indices, because the symmetry-related atoms have different values of x and y that are not related to one another by a translation. Higher-order n_m-screw axes parallel to z will similarly give zero structure factors unless l is a multiple of n/m.

The effect of glide planes is similar but leads to absences amongst reflections with one zero index, if the glide plane is perpendicular to an axis. Thus a c-glide plane perpendicular to the y-axis (but not necessarily through the origin) repeats an atom at x, y, z to $x, y', z+\tfrac{1}{2}$ so that

$$F_{hkl} = \sum_i f_i \{\cos 2\pi(hx_i + ky_i + lz_i) + \cos 2\pi(hx_i + ky'_i + l(z_i + \tfrac{1}{2}))\}$$

and

$$F_{h0l} = \sum_i f_i (\cos 2\pi(hx_i + lz_i) + \cos 2\pi(hx_i + l(z_i + \tfrac{1}{2})))\}$$

$$= 2\sum_i f_i \cos 2\pi(hx_i + lz_i) \text{ for } l \text{ even}$$

and zero for l odd.

An n-glide plane perpendicular to y would similarly lead to absent $h0l$ reflections for $h+l$ odd. Similar, though mathematically more complex, considerations apply if the glide plane is not perpendicular to an axis.

The various indications of the presence of screw axes and glide planes are given in Table 14.1. In looking for evidence of their presence it is important to look first for the systematic absences amongst the general reflections hkl that indicate centred lattices, because these will of course carry through to reflections having one or two zero indices, but glide planes and screw axes corresponding to these absences may not necessarily exist. Equally glide planes should be sought before screw axes, because absences amongst reflections with one zero index will carry through to those with two zero indices, although the corresponding screw axis may not be present.

The extent to which the space-group of a crystal can be determined uniquely from systematically absent reflections depends on how much other information is available. There are ninety-seven different possible combinations of systematic absences and if these are combined with a knowledge of the crystal system the number of uniquely determinable space-groups is thirty-eight, while the remaining 192 space-groups can be divided on the basis of the absences into sets containing from two to eight members. If the point group is known the number of uniquely determinable space groups rises to 186, and the remaining forty-four form twenty-two ambiguous pairs. Methods of resolving these remaining ambiguities are beyond the scope of this book.

TABLE 14.1

If there are no reflections of the stated type other than those satisfying the stated condition then the symmetry element may be deduced as shown unless the same condition is already imposed by the Bravais lattice or (in the case of a reflection with two zero indices) by a glide plane.

Type of reflection	Condition for reflection	Symmetry element		System in which applicable			
$hk0$	$h = 2n$	a	$\left.\begin{array}{l}\\\\\\\end{array}\right\} \perp z$	cubic,	tetrag.,	orthorh.	
	$k = 2n$	b		,,	,,	,,	
	$h + k = 2n$	n		,,	,,	,,	
	$h + k = 4n$	d		,,	–	,,	
$h0l$	$h = 2n$	a	$\left.\begin{array}{l}\\\\\\\end{array}\right\} \perp y$	cubic,	tetrag.,	orthorh.,	monocl.
	$l = 2n$	c		,,	,,	,,	,,
	$h + l = 2n$	n		,,	,,	,,	,,
	$h + l = 4n$	d		,,	,,	,,	–
$0kl$	$k = 2n$	b	$\left.\begin{array}{l}\\\\\\\end{array}\right\} \perp x$	cubic,	tetrag.,	orthorh.	
	$l = 2n$	c		,,	,,	,,	
	$k + l = 2n$	n		,,	,,	,,	
	$k + l = 4n$	d		,,	,,	,,	
$hhl, h\bar{h}l$	$l = 2n$	c	on 110	cubic,	tetrag.,		
	$2h + l = 4n$	d	and $1\bar{1}0$	,,	,,		
$hkh, hk\bar{h}$	$k = 2n$	c	on 101	cubic			
	$2h + k = 4n$	d	and $10\bar{1}$	,,			
$hkk, hk\bar{k}$	$h = 2n$	c	on 011	cubic			
	$h + 2k = 4n$	d	and $01\bar{1}$	,,			
$\{hh\overline{2h}l\}$	$l = 2n$	c	on $\{1\bar{1}00\}$	hexagonal			
$\{h\bar{h}0l\}$	$l = 2n$	c	on $\{11\bar{2}0\}$	hexagonal			
$h00$	$h = 2n$	2_1	$\left.\begin{array}{l}\\\\\\\end{array}\right\} \parallel x$	cubic,	tetrag.,	orthorh.	
		4_2		,,	–	–	
	$h = 4n$	4_1 or 4_3		,,	,,	–	
$0k0$	$k = 2n$	2_1	$\left.\begin{array}{l}\\\\\\\end{array}\right\} \parallel y$	cubic,	tetrag.,	orthorh.,	monocl.
		4_2		,,	–	–	–
	$k = 4n$	4_1 or 4_3		,,	–	–	–
$00l$	$l = 2n$	2_1	$\left.\begin{array}{l}\\\\\\\end{array}\right\} \parallel z$	cubic,		orthorth.	
		4_2		,,		tetrag.	
	$l = 4n$	4_1 or 4_3		,,		,,	
$000l$	$l = 2n$	6_3	$\left.\begin{array}{l}\\\\\\\end{array}\right\} \parallel z$	hexagonal			
	$l = 3n$	$3_1, 3_2, 6_2, 6_4$		,,		trigonal	
	$l = 6n$	$6_1, 6_5$		,,		–	

The representation of space-group symmetry

The symmetry elements of a space-group are usully represented in their mutual relationships on a projection of the unit cell down the z-axis. The symbols used for the symmetry elements are shown in Fig. 14.5 and are a logical extension of those used in stereograms in Part I. For symmetry elements that lie parallel to the plane of projection it is necessary to specify the height in the cell (as a fraction of c) at which they lie, but this is omitted if it is zero. As an example Fig. 14.10(a) shows such a projection of the orthorhombic space group $P2_1/b\ 2/c\ 2_1/n$ (the shortened symbol of which is $Pbcn$). Corresponding diagrams of all 230 space-groups are given in *International Tables for X-ray Crystallography*, Vol. 1 (Kynoch Press, 1952). The conventional orientation of the axes is with the origin at the top left, with the x-axis pointing down the page and the y-axis to the right, exactly as in the stereograms in Part I.

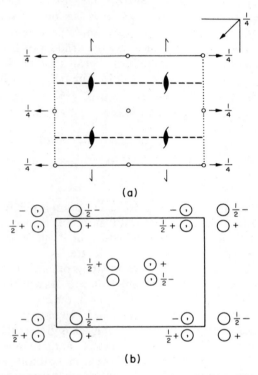

FIG. 14.10. (a) Arrangements of the symmetry elements, and (b) repetition of a point in the general position, in a unit cell of the space group $P2_1/b\ 2/c\ 2_1/n$, projected down the z-axis.

The effect of the symmetry elements is shown by inserting a point in a general position in the cell and then repeating it in accordance with the symmetry, in much the same way as in a sketch stereogram. This may be done either on the drawing of the symmetry elements, or, if necessary for clarity, it may be done on a separate drawing of the unit cell alongside that showing the symmetry elements. Figure 14.10(b) shows $P2_1/b\ 2/c\ 2_1/n$ again with such a set of points inserted. The initial point is at general coordinates x, y, z and is indicated by $\bigcirc +$, where the $+$ indicates an arbitrary positive z-coordinate. The z-coordinates of symmetrically related points are then at corresponding coordinates $-z, \frac{1}{2}+z$ and $\frac{1}{2}-z$ with the same value of z, and these are indicated by $-, \frac{1}{2}+$ and $\frac{1}{2}-$ respectively. Points that are related to the initial point by an inversion or reflection are denoted $\odot$; this indicates that if $\bigcirc$ has a "handed" environment then $\odot$ will have an environment of opposite hand.

It can be seen from Fig. 14.7 that in this space-group a point in a *general position* is repeated into eight positions. It is said to be an 8-fold position, or to have a multiplicity of 8. Its eight sets of symmetry-related coordinates are:

$$x, y, z; \quad \tfrac{1}{2}-x, \tfrac{1}{2}-y, \tfrac{1}{2}+z; \quad \tfrac{1}{2}+x, \tfrac{1}{2}-y, \bar{z}; \quad \bar{x}, y, \tfrac{1}{2}-z;$$
$$\bar{x}, \bar{y}, \bar{z}; \quad \tfrac{1}{2}+x, \tfrac{1}{2}+y, \tfrac{1}{2}-z; \quad \tfrac{1}{2}-x, \tfrac{1}{2}+y, z; \quad x, \bar{y}, \tfrac{1}{2}+z.$$

However, if the initial point lies on a mirror plane or at a centre of symmetry its multiplicity is reduced by a factor of 2, and if it lies on an n-fold rotation axis its multiplicity is reduced by a factor of n. If it lies simultaneously on more than one such symmetry element its multiplicity is reduced by the product of the appropriate factors. Points with reduced multiplicity are said to be in *special positions*. In Pbcn there are three different special positions, all of which are 4-fold. The point may lie on a 2-fold axis at $0, y, \tfrac{1}{4}$ and its set of four positions is then

$$0, y, \tfrac{1}{4}; \quad 0, \bar{y}, \tfrac{3}{4}; \quad \tfrac{1}{2}, \tfrac{1}{2}+y, \tfrac{1}{4}; \quad \tfrac{1}{2}, \tfrac{1}{2}-y, \tfrac{3}{4}.$$

Alternatively it may lie at a centre of symmetry with coordinates

$$0, \tfrac{1}{2}, 0; \quad 0, \tfrac{1}{2}, \tfrac{1}{2}; \quad \tfrac{1}{2}, 0, 0; \quad \tfrac{1}{2}, 0, \tfrac{1}{2};$$

or

$$0, 0, 0; \quad 0, 0, \tfrac{1}{2}; \quad \tfrac{1}{2}, \tfrac{1}{2}, 0; \quad \tfrac{1}{2}, \tfrac{1}{2}, \tfrac{1}{2}.$$

It is to be noted that there is in general no reduction of multiplicity if the point lies on a glide plane or a screw axis* because it is then still moved by the translation component of the symmetry operation; nor is there any reduction if the point lies on an axis of rotation–inversion unless it lies at the inversion centre.

In a crystal structure the number of equivalent atoms of particular kinds in the unit cell is decided by the multiplicity of the sites at which they lie. As will be seen in Chapter 15, if both the space-group and the number of like atoms in the cell is known this can be very helpful in deciding where these atoms must lie in order to have the correct multiplicity. If the number of atoms of a particular kind in the unit cell is greater than the multiplicity of the available sites in the space-group then they must lie at more than one set of sites. This can be very important in mineralogy because these atoms, although chemically identical, will then be crystallographically non-equivalent; their environments will be different, and they will be differently susceptible to isomorphous substitution by other elements. This may still be true even if the number of atoms is small enough to be accommodated on one set of sites but it is actually accommodated in two sets of lower multiplicity. For example, in the above space group there could be eight atoms of a particular element accommodated at the general position, or they could be divided into two sets of four and accommodated in two special positions. This distinction is very important in understanding the geochemical behaviour of a mineral.

The recognition of space-group symmetry

Crystal structure models often look extremely confusing, especially if the structure is at all complex, and it is difficult to see why they have so complicated a configuration. This can

---

* Except in the case of 4_2-, 6_2-, and 6_4- screw axes which as we have seen "contain" a 2-fold axis, and 6_3-axes which "contain" a 3-fold axis. These ordinary components have their usual effect in reducing multiplicity by a factor of 2 or 3 as the case may be.

Symmetry in Repeating Patterns 203

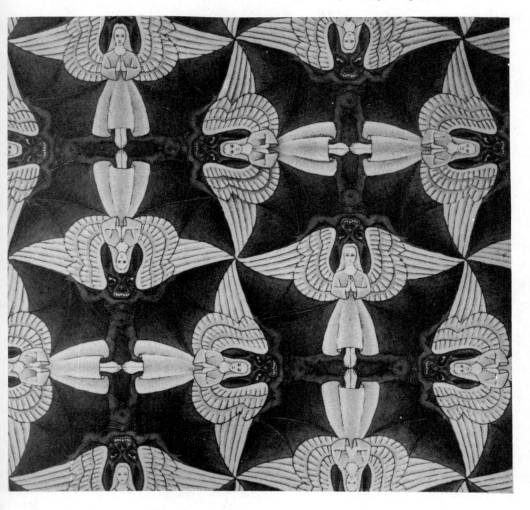

FIG. 14.11. A repeating pattern from the graphical work of M. C. Escher with a two-dimensional space group.

only be understood by recognising the positions and orientations of the symmetry elements in the structure, and practice in this can only be gained by handling structure models themselves; it cannot be conveyed by a written text or even from drawings or projections. However, some useful preliminary practice can be obtained by recognising two-dimensional space-groups and their symmetry elements in two-dimensional repeating patterns. These also have the advantage of much greater simplicity.

In two-dimensional repeating patterns there are only six kinds of symmetry elements: 2-fold, 3-fold, 4-fold and 6-fold rotation *points*, mirror *lines* and glide *lines*. The latter involve reflection in the line and translation of half a repeat distance parallel to it. In such patterns there are only four "crystal systems" (oblique, rectangular, square and

204   Crystallography for Earth Science Students

hexagonal), and there are five Bravais lattices (primitive in each system, and also centred in the rectangular system), ten point groups and seventeen "space" groups. Nomenclature is the same except that the lower-case letters *p* and *c* are used to denote primitive and centred Bravais lattices and *g* is used to denote a glide line. The arrangements of the symmetry elements in all seventeen two-dimensional space groups are illustrated in *International Tables for X-ray Crystallography*.

The recognition of space-group symmetry can therefore be practised on the repeating patterns on wallpaper, printed textiles, and floor coverings, and also *par excellence* on those in the artistic work of M. C. Escher.* One of these is reproduced in Fig. 14.11, and a corresponding drawing of the symmetry elements in the same orientation is shown in Fig. 14.12.

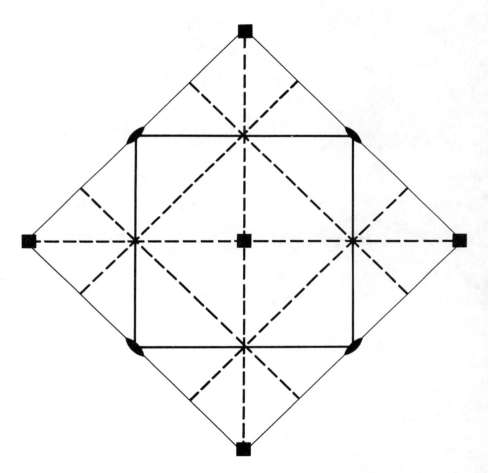

FIG. 14.12. The unit cell and symmetry elements of the pattern in Fig. 14.11, in the same orientation.

* M. C. Escher, *Graphical Work* (Oldbourne Press, London, 1961); see also C. H. Macgillavry, *Symmetry Aspects of M. C. Escher's Periodic Drawings* (Utrecht, 1965).

Problems

1. A monoclinic crystal is found to give reflections of the following types:
    hkl no restrictions,
    $hk0$ no restrictions,
    $h0l$ only with l even,
    $0kl$ no restrictions,
    $h00$ no restrictions,
    $0k0$ only with k even,
    $00l$ only with l even.
   What is its space group?
2. Another monoclinic crystal, known to belong to the holosymmetric class, is found to give reflections of the following types:
    hkl only with $h+k$ even,
    $hk0$ only with $h+k$ even,
    $h0l$ only with h even and l even,
    $0kl$ only with k even,
    $h00$ only with h even,
    $0k0$ only with k even,
    $00l$ only with l even.
   What is its space group?
3. A monoclinic crystal, known from its piezo-electric properties to have no centre of symmetry, gives reflections of all types with no restrictions of indices. To what space-groups may it belong?

CHAPTER 15

The Determination of Crystal Structures

IF THE structure of a crystal of a complicated composition is totally unknown its elucidation can be a long and complex process. The crystal will usually be investigated first of all by taking single-crystal photographs to find the unit cell and space group by the kind of methods that we have already discussed. It will then be necessary to determine the relative intensities of several thousand individual reflections, which these days will usually be done on a computer-controlled single-crystal diffractometer, whose design and operation it is unnecessary to go into here. One then has to find an approximation to the structure. This may be achieved in a variety of ways, depending on the complexity of the problem. It will certainly involve consideration of the chemistry, the number of atoms of each element in the unit cell, and the way these can be fitted into a structure consistent with the space-group. Physical properties, model building and analogies with known structures may help, as also may considerations of the magnitude of some of the largest structure factors. By whatever combination of methods proves necessary, a trial structure is eventually formulated which assigns coordinates to each atom in the unit cell and so enables one to compute the structure factors for all the thousand or more reflections. These calculated structure factors (F_{calc}) are then compared with the observed ones (F_{obs}) by calculating a "residual" (or discrepancy factor) such as

$$R = \frac{\sum_{hkl} ||F_{obs}| - |F_{calc}||}{\sum_{hkl} |F_{obs}|}$$

where the summations are taken over all reflections hkl. The agreement does not have to be very good at this stage. If R is less than about 0.4 it is probably sufficiently good to permit the structure to be "refined". This involves very extensive computations to find changes in the atomic coordinates that minimise the discrepancies between the values of $|F_{calc}|$ and $|F_{obs}|$ and so reduce R. Eventually it proves impracticable to reduce R any further, which usually happens when it is in the range 0.03 to 0.06. At this stage the atomic positions will probably be correct to within a few thousandths of an Å. If the attempted refinement fails to reduce R substantially it is an indication that the trial structure was wrong and the initial moderate agreement between $|F_{calc}|$ and $|F_{obs}|$ was fortuitous. A new trial structure must then be sought.

A great deal of structural work in modern mineralogy is not concerned with totally unknown structures, however, since the crystal structures of all the main mineral groups have already been determined. Rather it is concerned with the exact details of interatomic distances, and especially the relative degrees of occupation of the sites in the structures by atoms of different elements as a function of changes of composition (or even merely of temperature and pressure) in a given mineral type. The initial experimental work and the computational work of refining the structure are the same as has been described, but the problem of finding a trial structure does not exist: a previously determined structure of another member of the same mineral group will be amply good enough for the purpose.

This book does not attempt to provide the student with the expertise required for either of these kinds of research, but in the remainder of this chapter we shall exemplify the principles involved by showing how a very simple crystal structure, that of rutile, can be determined from scratch, and how the sites occupied by different cations in a solid solution series can be elucidated.

Determination of the structure of rutile

Rutile (TiO_2) occurs in tetragonal crystals whose morphology shows that they belong to class $4/m\,2/m\,2/m$. A rotation photograph about the tetrad axis gives a value of $c = 2.96$ Å from the layer line repeat, and indexing of the zero layer line by the methods described in Chapter 13 permits one to determine $a = 4.59$ Å from the same photograph.

The next step in the investigation is to find out how many atoms of each kind there are in the unit cell. To do this it is necessary to measure the density of the crystal, either by Archimedes' method, if a large enough crystal is available, or by immersion in a heavy liquid whose density can be adjusted until the crystal neither sinks nor floats. In the case of rutile the density is found to be about 4.25 g cm^{-3}, and the volume of the unit cell is given by

$$V = a^2 c = 62.4 \text{ Å}^3$$
$$= 62.4 \times 10^{-24} \text{ cm}^3.$$

Thus the mass of the unit cell is 265×10^{-24} g. This mass can be converted from grams to atomic weight units by multiplication by Avogadro's number, 0.602×10^{24}. The unit cell therefore contains 159.5 atomic weight units, and since the formula weight of $TiO_2 = 47.9 + 2 \times 16 = 79.9$ it follows that there must be two formula units in the unit cell, i.e. 2 atoms of titanium and 4 atoms of oxygen.

Determination of the space-group is the next step. Because the rotation photograph about the z-axis does not enable us to distinguish between reflections 430 and 500, and does not permit reflections of type $00l$ to be recorded, we would probably decide to take additional oscillation photographs about both the x- and z-axes and to index them using the reciprocal lattice construction. From such evidence it is found that:

hkl reflections are present with $h+k+l$ even or odd, so the Bravais lattice is P.

$hk0$ reflections occur with h, k and $h+k$ even or odd, showing that the symmetry plane perpendicular to z must be an ordinary mirror plane and not a glide plane.

$0kl$ (and also of course by the tetragonal symmetry $h0l$ reflections) occur only with $k+l$ even, showing that the plane of symmetry perpendicular to x (and y) is an n-glide plane.

hhl reflections are present with l even or odd, showing that the diagonal symmetry plane is a mirror plane.

208   Crystallography for Earth Science Students

Reflections of types $h00$, $0k0$ and $00l$ are absent with the non-zero index odd, but these absences do not provide any additional information because they arise necessarily as special cases of the absences among $0kl$ (and $h0l$) reflections. However, reference to the *International Tables for X-ray Crystallography* will show that there is only one space group that satisfies the conditions that we have found, and this is $p4_2/m\ 2_1/n\ 2/m$ ($P4_2/mnm$ in contracted notation).

The space-group is of course fully illustrated in the *International Tables*, but it will be instructive to build up a diagrammatic representation of it in stages. Figure 15.1(a) shows two unit cells, into which have been inserted the planes of symmetry found from diffraction evidence. It is to be noted that planes of symmetry in repeating patterns are always separated by half a cell, not a whole cell, and either lie at 0 and $\frac{1}{2}$ or at $\frac{1}{4}$ and $\frac{3}{4}$. The origin is taken at the intersection of three mirror planes (the two diagonal ones and that perpendicular to the z-axis) for the sake of convenience, and it can be shown that the only self-consistent positions for the glide planes are then at $\frac{1}{4}$ and $\frac{3}{4}$.

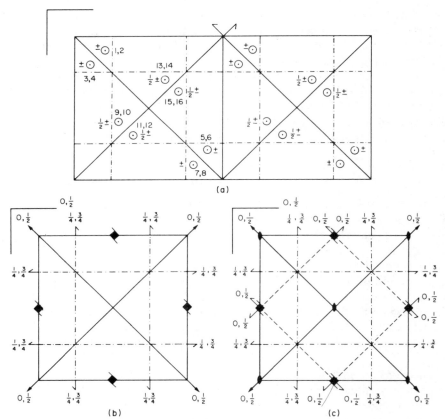

FIG. 15.1. (a) Two tetragonal unit cells with the symmetry planes found from the diffraction results from rutile; a point (1) at x, y, z repeated sixteen-fold by this symmetry. (b) Symmetry as in (a) plus axes of symmetry deduced from (a) that are specified in the symbol $P4_2/m\ 2_1/n\ 2/m$. (c) Further symmetry elements deduced from (a) inserted to give the full representation of the space group.

If we insert a general point at x, y, z, and repeat it by the symmetry elements we have already placed, the whole symmetry will become clear. The point marked 1 is the starting-point. It is repeated to 2 at $x, y, \bar{z}$ by the m-plane perpendicular to z, and points 1 and 2 are repeated to 3 and 4 by one of the diagonal mirrors. These four points are then repeated to 5, 6, 7, 8 by the other diagonal mirror. Point 1 is repeated by the n-glide at $y = \frac{1}{4}$ to position 9 at $\frac{1}{2}+x, \frac{1}{2}-y, \frac{1}{2}+z$. Positions 10–16 can then be generated from 9 using the diagonal mirrors and the mirror perpendicular to z at $z = \frac{1}{2}$, or they can be generated from 2–8 by glide planes. Further operations of any of the symmetry elements lead back to previously found positions.

At this stage the presence of any kind of tetrad is far from obvious, but if the corresponding points are inserted in the adjacent cell it becomes clear. Using primes to denote corresponding points in the next cell, we can see that points 3', 9', 5 and 15 lie (in projection) at the corners of a square centred at $x = \frac{1}{2}$, $y = 1$ and are respectively at z-coordinates of $z, z+\frac{1}{2}, z, z+\frac{1}{2}$. They are therefore related by a 4_2-screw axis, and so such axes may be inserted at the mid-points of the cell edges as in Fig. 15.1(b). There is no pair of points related by a 2-fold rotation axis parallel to x or y, but there are 2_1-screw axes in these directions: for example point 1 is related to 10 by a 2_1-screw axis at $y = \frac{1}{4}, z = \frac{1}{4}$. There are diagonal 2-fold rotation axes: for example, 9 is related to 12 by a diagonal 2-fold axis at $z = \frac{1}{2}$. Thus the nature of the axes of symmetry in the space-group symbol are found to be $4_2, 2_1$ and 2, respectively, even though we were unable to be sure of this directly from the absent reflections.

In Fig. 15.1(b) all the symmetry elements found above, and specified in the full space-group symbol, have been inserted. There are, however, still further symmetry elements automatically implied by these which can be found from a consideration of Fig. 15.1(a). There are: diagonal glide planes with a glide of $\frac{1}{2}a + \frac{1}{2}b$ interleaving between the diagonal mirror planes and relating point 13 to 9' for example; diagonal 2_1-screw axes at $z = 0$ and $\frac{1}{2}$ on these planes; 2-fold axes at the corners and centre of the cell; and centres of inversion at the corners of the cell, the centre of the cell, the mid-points of the edges, and the centres of the faces. Because these are all implicit in the others they are not specified even in the full notation of the space-group, but they are shown (except for the centres of inversion) in a full diagrammatic representation of it in Fig. 15.1(c).

It is evident from Fig. 15.1(a) that the general position in the space-group $P4_2/m\ 2_1/n\ 2/m$ is 16-fold, but we have already found that there are only two titanium atoms and four oxygen atoms in the unit cell. It follows that all the atoms lie in special positions with reduced multiplicity.

For the Ti atoms to lie in a position with a mulplicity of only 2, it is evident that they must be at the intersection of three mirror planes. If one of the Ti atoms is taken to lie at the origin 0, 0, 0 the symmetrically related point is at the centre of the cell, $\frac{1}{2}, \frac{1}{2}, \frac{1}{2}$. We do not need to consider another possibility, which would be to put the first atom at 0, 0, $\frac{1}{2}$, because it could be converted into the former simply by shifting the origin of the cell by $c/2$, so the two possibilities are equivalent to one another. The location of the Ti atoms in the structure is thus uniquely defined merely by consideration of the space group.

The oxygen atoms must be at a 4-fold site, and there are five different ways in which such a site can be derived by moving the general position on to symmetry elements. They are shown in Fig. 15.2 and described in (i)–(v) below.

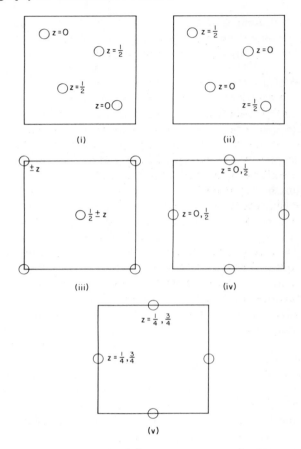

Fig. 15.2. The five sets of 4-fold positions in space group $P4_2/m\,2_1/n\,2/m$.

(i) On one diagonal mirror plane and on the mirror plane at $z = 0$. The four positions are $x, x, 0$; $\bar{x}, \bar{x}, 0$; $\frac{1}{2}+x, \frac{1}{2}-x, \frac{1}{2}$; $\frac{1}{2}-x, \frac{1}{2}+x, \frac{1}{2}$.

(ii) As (i) but on the mirror plane at $z = \frac{1}{2}$. The four positions are $x, x, \frac{1}{2}$; $\bar{x}, \bar{x}, \frac{1}{2}$; $\frac{1}{2}+x, \frac{1}{2}-x, 0$; $\frac{1}{2}-x, \frac{1}{2}+x, 0$.

(iii) On both diagonal mirror planes at $0, 0, z$; $0, 0, \bar{z}$; $\frac{1}{2}, \frac{1}{2}, \frac{1}{2}+z$; $\frac{1}{2}, \frac{1}{2}, \frac{1}{2}-z$.

(iv) At the intersection of the 4_2-screw axis with the mirror plane at $z = 0$ giving the four positions, $0, \frac{1}{2}, 0$; $\frac{1}{2}, 0, 0$; $0, \frac{1}{2}, \frac{1}{2}$; $\frac{1}{2}, 0, \frac{1}{2}$.

(v) On the 4_2-screw axis (whose 2-fold axis component reduces the multiplicity by 2) at $z = \frac{1}{4}$ so that the repetition by the 4_2-axis coincides with the effect of reflection in the mirror planes at $z = 0$ and $\frac{1}{2}$. The four positions are $0, \frac{1}{2}, \frac{1}{4}$; $\frac{1}{2}, 0, \frac{1}{4}$; $0, \frac{1}{2}, \frac{3}{4}$; $\frac{1}{2}, 0, \frac{3}{4}$.

A choice may be made between these positions on the evidence of the intensities of the X-ray reflections. If we substitute the Ti coordinates into the structure factor equation we

The Determination of Crystal Structures 211

get the contribution of the Ti atoms to any reflection, as

$$F(\text{Ti})_{hkl} = f_{\text{Ti}} \cos 2\pi(h.0 + k.0 + l.0) + f_{\text{Ti}} \cos 2\pi(h.\tfrac{1}{2} + k.\tfrac{1}{2} + l.\tfrac{1}{2})$$
$$= 2f_{\text{Ti}} \text{ for } h + k + l \text{ even}$$
and zero for $h + k + l$ odd.

Thus the Ti atoms do not contribute to F_{hkl} with $h + k + l$ odd, and such structure factors therefore depend only on the oxygen positions. For any of the sets of sites (iii)–(v) substitution of the four sets of coordinates into the structure factor equation shows that atoms in these sites would also make a zero contribution to the structure factor for $h + k + l$ odd. Since the intensities of reflections with indices of this type are not zero (as we saw in deducing the Bravais lattice) it follows that sites (iii)–(v) can be ruled out. Furthermore, as can be seen from Fig. 15.2, (ii) is equivalent to (i) if we rotate the cell through 90° and change the (so far arbitrary) value of x. Only sites (i) need therefore be considered for the oxygen atoms, and we are left with a single unknown parameter, x, to find.

A first approach to this final problem is "model building", though the necessary work can be done on paper without actually constructing a three-dimensional model. All the atoms lie on one or other of the two diagonal planes $1\bar{1}0$ and 110, and sections, on these two planes, of one unit cell of the structure are shown in Fig. 15.3. It is clear that each Ti atom is surrounded by six oxygen atoms, and four of these lie in one plane and are equidistant from the Ti atom, and two are collinear with the Ti on the perpendicular plane.

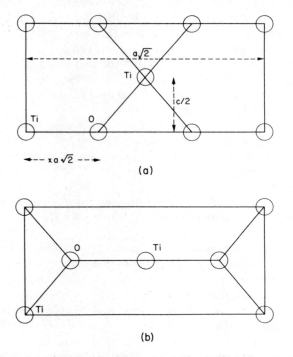

FIG. 15.3. Diagonal cross-section of the unit cell of rutile parallel to (a) the $1\bar{1}0$ plane and (b) the 110 plane. The coordinate of oxygen is chosen to make all Ti–O distances equal.

The latter two are also equidistant from Ti, but not necessarily at the same distance as the first four. Each oxygen lies in the latter relationship to one Ti atom and in the former relationship to two. A very reasonable model would be to suppose all six Ti–O distances to be equal. From Fig. 15.3 it can be deduced that this requires

$$ax\sqrt{2} = \{2a^2(\tfrac{1}{2}-x)^2 + c^2/4\}^{1/2},$$

whence
$$x = \tfrac{1}{4} + \tfrac{1}{8}c^2/a^2$$
$$= 0.30.$$

The equality of the Ti–O distances is of course only a hypothesis and must be confirmed, and if necessary refined, by a comparison of observed and calculated structure factors. Because we have only one variable to consider in this simple structure the process is very easy. Two structure factors, say F_{111} and F_{210}, to which the Ti atoms do not contribute, may be calculated for three values of x, say 0.28, 0.31 and 0.34, by inserting these values in

$$F_{111} = f_O(111)\{\cos 2\pi(x+x+0) + \cos 2\pi(-x-x+0) + \cos 2\pi(\tfrac{1}{2}+x+\tfrac{1}{2}-x+\tfrac{1}{2})$$
$$+ \cos 2\pi(\tfrac{1}{2}-x+\tfrac{1}{2}+x+\tfrac{1}{2})\}$$

and $F_{210} = f_O(210)\{\cos 2\pi(2x+x) + \cos 2\pi(-2x-x) + \cos 2\pi(1+2x+\tfrac{1}{2}-x)$
$$+ \cos 2\pi(1-2x+\tfrac{1}{2}+x)\},$$

where $f_O(111)$ and $f_O(210)$ are the values of the atomic scattering factor of oxygen for the values of $\sin\theta/\lambda$ corresponding to the 111 and 210 reflections respectively (see Chapter 10). The ratio $|F_{111}|/|F_{210}|$ can then be plotted against x. The observed values of $|F_{111}|$ and $|F_{210}|$ can be found by correcting the observed intensities of the corresponding reflections for the various experimental factors (Lorentz, polarisation, etc.) and taking their square roots. The value of x for which $(|F_{111}|/|F_{210}|)_{calc} = (|F_{111}|/|F_{210}|)_{obs}$ can therefore be found. It is in fact 0.305, so that the two kinds of Ti–O distance in the structure are nearly, but not quite, equal as we expected. They are actually 1.95 and 1.98 Å.

The process of determining the structure of rutile exemplifies the basic features involved in all crystal structure determinations, namely:

(i) measurement of the unit cell;
(ii) measurement of the density, and hence calculation of the numbers of atoms of each kind in the unit cell;
(iii) determination of the space-group;
(iv) finding appropriate kinds of site in the space-group for each kind of atom, using information from the diffraction pattern where possible, and culminating in a trial model;
(v) refining the variable parameters in the model (that are not fixed by the space-group), by comparison of calculated and observed structure factors.

In rutile the problem is simplified because of the small number of atoms in the unit cell and the fact that they are in special positions, and as a result there is only one parameter to refine. In more complicated structures there may be of the order of 100 atoms in the unit cell, mainly in general positions with three variable parameters x, y and z. There are therefore likely to be tens of parameters to refine, requiring extensive least-squares computation to fit large numbers of observed and calculated structure factors. The fundamental principles, however, remain the same.

Refinement of the structure of solid solutions

It is comparatively rare for minerals to be stoichiometric compounds. They usually have variable amounts of substitution of one element by another, as for example in enstatite which may well have a formula such as $Mg_{0.72}Fe_{0.28}SiO_3$. The unit cell of enstatite contains sixteen formula units, but any particular unit cell clearly cannot contain 11.52 Mg atoms and 4.48 Fe atoms – these can only be average figures. There must therefore be some irregularity in the crystal structure, the contents of different unit cells being different; some will probably contain 12Mg+4Fe, others 11Mg+5Fe, and others may depart even further from the mean. This irregularity means that diffracted intensity will not be confined entirely to reflections at the Bragg angles, because the derivation of the Bragg law assumed complete regularity throughout the structure. Fortunately, however, the structure can be regarded as the sum of two parts, the average structure which repeats regularly, and a "difference structure" containing more or less random departures (positive and negative) from the average structure. It can be shown that each of these components gives its own diffraction effects superimposed on those from the other. The diffraction pattern due to the average structure will be at the Bragg angles in the usual way, while that from the "difference structure" will be spread over all possible angles and so is everywhere very weak, mainly inappreciable, and certainly not seriously modifying the intensities of the Bragg reflections.

In the structure factor expression

$$F_{hkl} = \sum_i f_i \cos 2\pi(hx_i + ky_i + lz_i)$$

we can therefore consider the summation to be over all the i sites in the cell regardless of whether they happen to be occupied by (in the case of enstatite) Mg or Fe, provided that the scattering factor associated with those sites is taken as the weighted mean value

$$f_i = nf_{Mg} + (1-n)f_{Fe}$$

where n is the fraction of the ith sites occupied on average by Mg.

In the space-group of enstatite ($Pbca$) the multiplicity of the general position is 8-fold. Thus the sixteen metal atoms must lie in two crystallographically unrelated sets of sites, having different structural environments. There is therefore no reason why these two sets of sites (referred to as M1 and M2 for convenience) should be equally appropriate for occupation by Mg and Fe, and it is entirely possible that the average structure should contain a preponderance of the Fe in M1 and rather little in M2 or vice versa. In fact the latter is what happens. The scattering factors for the two sites are therefore

$$f_{M1} = n_1 f_{Mg} + (1-n_1)f_{Fe}$$
and
$$f_{M2} = n_2 f_{Mg} + (1-n_2)f_{Fe}$$

where $(n_1 + n_2)/2 = n$, the overall fraction of Mg in the specimen as a whole.

When different values of f_{M1} and f_{M2} are respectively associated in the structure factor expression with the different coordinates of the M1 and M2 sites the calculated value of any particular structure factor will depend on the extent of the segregation of Fe into M2. The value of n_1 (or $n_2 = 2n - n_1$) is therefore an additional variable parameter which can (and must) be adjusted in matching the F_{calc} values to the F_{obs} values. Such parameters are described as *site occupancies*, and their determination is a very important feature of modern crystal structural investigations of minerals because they reveal the *degree of ordering* of elements into different sites. This depends in part on the relative bonding energies of the atoms in the different sites, but also on the temperature of formation of the mineral. At its formation the crystal will have adopted a configuration that gave rise to the minimum possible value of its free energy G, where

$$G = H - TS.$$

Complete ordering of the atoms into their most strongly bonded sites minimises the enthalpy H, but disordering increases the entropy S and so may diminish G even though increasing H. The degree of ordering that is adopted takes account of these opposing effects, and the balance between them clearly depends on the temperature T.

The accuracy with which site occupancies can be found depends on the sensitivity of the weighted mean atomic scattering factor to the weights n_1 and $1 - n_1$, and this depends on the difference between the scattering factors of the atoms concerned. The scattering factors of Mg and Fe are very different, and the site occupancies of these elements in enstatite can be determined to an accuracy of about 0.01. This is because Mg and Fe differ substantially in atomic number, and the scattering factors of atoms are proportional to atomic number at $\sin\theta/\lambda = 0$ and diminish with increasing $\sin\theta/\lambda$ in similar (though not quite identical) ways, as shown in Fig. 10.2. For elements that are close to one another in atomic number like Al and Si, or Mn and Fe, occupancies cannot be satisfactorily determined in this way because of the similarity of their scattering factors. Relative values of f for a range of important elements are shown (at a particular value of $\sin\theta/\lambda$) in Fig. 15.4.

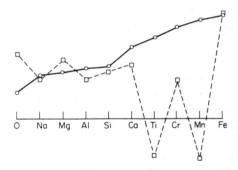

FIG. 15.4. Comparison of element-to-element variation of scattering factor for neutrons (□) with that for X-rays at $\sin\theta/\lambda = 0.3$ (○). The scale of the neutron data is expanded by a factor of 5 relative to the X-ray data. Some neutron values are negative.

In some circumstances this problem can be overcome by using neutron diffraction instead of X-ray diffraction. Although neutrons are normally thought of as particles, quantum effects involve their having a wave-like nature with a wavelength dependent on their velocity according to the relation

$$\lambda = h/mv$$

where h is Plank's constant, m the mass of the neutron and v its velocity. In numerical terms this works out to

$$\lambda = \frac{375}{v} \text{ Å}$$

if v is expressed in metres/sec. Neutrons that have undergone many atomic collisions in passing through a graphite or heavy water moderator in an atomic reactor have a distribution of velocities like the molecules of a gas at the same temperature, and at 100°C the root mean square velocity of the emerging neutrons is such as to correspond to a wavelength of about 1.3 Å, just in the right range for crystal diffraction. A narrow wavelength band can be isolated by reflection of the neutrons at the appropriate Bragg angle from a crystal plate to give an effectively monochromatic neutron beam. The principles of neutron diffraction with such a beam are entirely analogous to those of X-ray diffraction.

Neutrons have a number of disadvantages compared with X-rays. The equipment to generate them is many orders of magnitude more expensive, the beams produced are relatively weak, and they interact much more weakly with matter with the result that much bigger specimens have to be used to get measurable effects. But they have one big advantage in that they interact with the nuclei of atoms instead of with the electrons, and the scattering powers of nuclei for neutrons vary enormously from one element to another; in fact they even vary substantially between different isotopes of the same element. These variations are evident in Fig. 15.4, from which it may be seen that strong contrast can be obtained between Mn and Fe by neutron diffraction whereas these elements are scarcely distinguishable on the basis of their X-ray scattering factor. Mg and Al also show a reasonable contrast, though Al and Si are still virtually indistinguishable.

Another advantage of neutron diffraction over X-ray diffraction is in locating hydrogen atoms in crystal structures. Because hydrogen atoms contain so few electrons their scattering factor often exerts a negligible effect on X-ray structure factors, and the positions of such atoms then cannot be found by X-ray crystallography. However, neutron scattering by hydrogen atoms, especially if they are substituted by the deuterium isotope, is substantial and affords a means of locating them.

As has been noted above, Al and Si have such similar scattering factors for both X-rays and neutrons that they are indistinguishable. Since Al substitution for Si in silicates is a very common and important phenomenon it is fortunate that another approach to Al, Si occupancy is possible. The two atoms are of slightly different sizes, so that the four oxygen atoms bonded to an Al atom are slightly further away from it than they would be from an Si atom, the appropriate distances being 1.75 Å and 1.61 Å respectively (subject to still smaller variations as a result of other factors). If a set of tetrahedral sites, T, in a crystal structure is occupied randomly by Al and Si in the ratio $n : 1 - n$, then the atomic positions that are found by refining the structure will correspond to a T–O distance that is the weighted mean of the distances appropriate to Al–O and Si–O. Thus the Al occupancy in a tetrahedral site in a silicate can be found from

$$n = \frac{(\text{T–O dist.}) - 1.61}{1.75 - 1.61}.$$

Similar methods can also be applied to other occupancy problems as an additional or alternative method to that based on scattering factors.

Problems

1. Brucite is trigonal, and consists of layers, perpendicular to the z-axis, in which the Mg atoms are coplanar and are symmetrically sandwiched between planes of OH at either side. The 0001 reflection is strong but the 0002 and 0003 reflections are of approximately zero intensity. Find the distance between the planes of oxygen atoms given that:

$$d_{0001} = 4.77 \text{ Å}.$$

The values of f_{Mg} for the three reflections 0001, 0002 and 0003 are 9.6, 8.6 and 7.4 and those of f_O are 7.7, 5.5 and 3.9. The scattering factor of hydrogen is negligibly small.

2. The spinel (XY_2O_4) structure has atoms at the following positions in the unit cell:
   (a) Tetrahedral cations: $\frac{1}{8}, \frac{1}{8}, \frac{1}{8}$; $\frac{7}{8}, \frac{7}{8}, \frac{7}{8}$.
   (b) Octahedral cations: $\frac{1}{2}, \frac{1}{2}, \frac{1}{2}$; $\frac{1}{2}, \frac{1}{4}, \frac{1}{4}$; $\frac{1}{4}, \frac{1}{2}, \frac{1}{4}$; $\frac{1}{4}, \frac{1}{4}, \frac{1}{2}$.
   (c) Oxygen: x, x, x; $x, \frac{1}{4}-x, \frac{1}{4}-x$; $\frac{1}{4}-x, x, \frac{1}{4}-x$; $\frac{1}{4}-x, \frac{1}{4}-x, x$;

   $\bar{x}, \bar{x}, \bar{x}$; $\bar{x}, \frac{3}{4}+x, \frac{3}{4}+x$; $\frac{3}{4}+x, \bar{x}, \frac{3}{4}+x$; $\frac{3}{4}+x, \frac{3}{4}+x, \bar{x}$.

In normal spinels the X atoms are at (a) and the Y atoms at (b); in inverse spinels half the Y atoms are at (a) and the other half together with the X-atoms are at (b). Spinel itself, $MgAl_2O_4$, is found to have structure factors for neutron reflections from the 444 and 888 planes in the ratio

$$|F_{444}|/|F_{888}| = 0.86.$$

The neutron scattering factors (at all angles) for Mg, Al and O are 0.54, 0.35 and 0.58, and in this particular spinel the oxygen positions are given by $x = 0.262$.

Determine whether spinel is normal or inverse.

CHAPTER 16

Electron Diffraction in the Electron Microscope

Principles of the method.

THE wave-like nature of electrons in motion leads to the possibility of diffraction of electron beams by crystals. In principle this is similar to the diffraction of X-rays and neutrons, but there are major differences in practice because electrons interact with matter much more strongly than X-rays (whereas neutrons interact much less strongly), and the available wavelengths are much shorter. The strong interaction between electrons and matter is a complicating factor, because it means that a significant fraction of an incident electron beam is reflected by a single slab of crystal structure between two adjacent members of a set of lattice planes; thus the beam is seriously attenuated in its passage through the crystal, and reflections at successive planes after the first will be progressively weaker, whereas in the simple theory of crystal diffraction discussed in this book they have all been implicitly assumed to be equal. A further problem is that a reflected ray on its way out through the crystal may be reflected by another set of lattice planes that lies in an appropriate orientation, thereby attenuating the first reflected ray and giving rise to another (doubly reflected) ray in another direction. Such effects make electron diffraction impracticable if the crystals are more than a few hundred Ångstrom units thick. Fortunately electrons have a compensating advantage over X-rays in that they can readily be focused by appropriately arranged electric and magnetic fields (which is what makes the electron microscope possible), and there is no problem in viewing and manipulating very small crystals of appropriate size for electron diffraction in the electron microscope. Electron diffraction in the electron microscope is therefore a very appropriate technique for the study of the fine-grained materials that are common in petrological work, and it is becoming of increasing importance in this field.

A second difference between electron diffraction on the one hand and X-ray and neutron diffraction on the other is the wavelength that is employed. The wavelength of an electron depends on its velocity, as in the case of neutrons, and the velocity in its turn depends on the voltage used to accelerate the electron. The relationship is much more complicated than that given in Chapter 15 for neutrons because, in order to penetrate even the very thin specimens used, the electrons have to be accelerated to high velocities approaching that of light and relativistic effects are important. Some electron wavelengths at corresponding voltages are

50 kV	0.0535 Å
100 kV	0.0370 Å
500 kV	0.0142 Å
1000 kV	0.0087 Å

Thus the wavelengths available are of the order of 1/100 of those that we have regarded as suitable in X-ray and neutron diffraction. This means that the angles of diffraction obtained will also be of the order of 1/100 of those to which we are accustomed, and that all the rays involved in producing the whole diffraction pattern of a crystal will lie within about 1° of the undeviated beam. This would be a hopeless situation with X-rays, but the possibility of focusing electron beams again comes to our rescue; the beam can be focused so finely that very close diffraction spots are resolved from one another, and a magnified image of the diffraction pattern can be produced by the electron microscope so that the pattern can be made easily visible and can be photographed on a reasonable scale.

The principle of the method is shown in Fig. 16.1, which shows a parallel incident beam of electrons hitting the specimen and being diffracted by it. (The electron gun and system of condenser lenses required to produce this beam are omitted for clarity.) The diffracted rays are collected by the objective lens (L) of the electron microscope (drawn as though it were an ordinary optical lens) which can be regarded as focusing them in two different senses. All the electrons diffracted from each particular point of the object are focused to a corresponding point in the plane I, so that in this position there is formed the primary image of the object. On the other hand, all the electrons diffracted in a given direction from

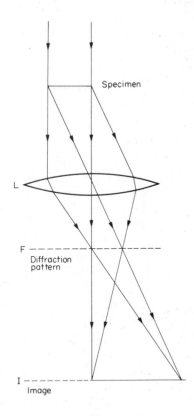

FIG. 16.1. Ray diagram showing the focusing of an electron diffraction pattern by the objective of an electron microscope.

all parts of the object (i.e. all the electrons in a particular diffracted ray corresponding to a particular set of crystallographic planes) are travelling in a parallel beam after diffraction (just as they were before diffraction), and they therefore come to a focus in the focal plane of the objective, F. Thus each diffraction direction will be represented by a sharp spot in this plane. The electron microscope contains a further system of projection lenses which normally serve to magnify the primary image I on to a fluorescent screen (or photographic film); thus, if this projection system is focused on to F instead of I it produces a magnified image of the diffraction pattern on the screen (or photographic film) instead of a magnified image of the object.

Some practical details

Specimens for electron diffraction and transmission electron microscopy are commonly mounted on small mesh grids. Such a grid may then be covered with a thin amorphous carbon film, about 100 Å thick, prepared by evaporating carbon *in vacuo* on to a clean smooth surface from which it may be floated off on to the surface of water and then picked up on the copper grid. A refinement is to use a "holey" carbon film, having holes in it a few hundred Å in diameter, so that specimen particles can be supported by bridging across these holes and so can be examined with very little superimposition even on the thin carbon film. Particulate specimens are prepared by very fine grinding and ultrasonic dispersion of the powder in water or some other liquid to give an exceedingly dilute suspension. A single drop of this suspension is then placed on the carbon film and evaporated to dryness, and with luck will be found to have deposited crystal fragments suitable for examination.

Another method of specimen preparation which is of increasing importance in mineralogy and petrology starts from a petrographic thin section of, say, 30 μm thickness. A region of interest in this can be chosen in the optical microscope, and this region is then thinned by bombardment from both sides with argon ions *in vacuo*. The process is continued until a hole is produced, and the thin wedge-shaped region round the hole then contains crystals of suitable thickness for examination. This method is very much more time-consuming than the liquid suspension method, but it has the advantage that the orientation of the crystals is preserved relative to their surroundings.

The specimen, of whatever type, is held on a specimen stage which permits it to be traversed in two directions across the axis of the microscope and also tilted with respect to the beam. The traverse is used to find a particle of interest and bring it to the centre of the field, using the instrument in microscope mode; after switching over to diffraction mode the diffraction pattern of the particle can be seen and the specimen can be tilted to enable different diffraction effects to be observed. However, since the electron beam will usually be much wider than one particular particle and will be diffracted by all particles on which it is incident, it is necessary to take special steps to isolate the diffraction pattern of the chosen particle – to obtain what is called a "selected area diffraction pattern". This is done by inserting a metal plate at I in Fig. 16.1 with a small aperture in it. If this aperture only transmits those rays which form an image at I of the chosen particle, then when the projector lenses are focused on F they will produce a magnified image only of those features of the diffraction pattern at F that arise from the chosen particle, because all other rays have been intercepted at I. If the primary image at I is magnified, say, 50 times then a 50-μm aperture at I will isolate the diffraction effects from a selected area of the specimen 1 μm in diameter.

The interpretation of the diffraction pattern

Providing that the diffracting crystal is very thin we can to a first approximation ignore the complications that arise from the strong interaction of the electrons with the crystal, and treat the process in the same way as X-ray diffraction, in terms of the reciprocal lattice of the crystal. If a point *hkl* of the reciprocal lattice lies on the sphere of reflection, the Bragg Equation will be satisfied for the set of planes *hkl*, and the corresponding reflection will be produced. However, because the crystal is very small, the number of planes in the set is small and an appreciable departure from the ideal Bragg angle will be permitted (see Chapter 9 for a discussion of this effect for X-rays). This latitude can be regarded as equivalent to some degree of smearing of the reciprocal lattice points into finite spots, and there is therefore an increased probability that the sphere of reflection will intersect some of these spots even though the crystal is in an arbitrary orientation.

There is another very important difference from the X-ray case that arises from the short wavelength of electrons. In interpreting X-ray photographs we have $\lambda \sim 1.5$ Å, and if the unit cell measures about 10 Å then the reciprocal lattice parameters (a^*, b^*, c^*) are of the order of 0.15 reciprocal lattice unit, and the sphere of reflection is defined as having a radius of 1 reciprocal lattice unit. If we proceed in the same way with electrons having $\lambda = 0.03$ Å the reciprocal lattice parameters will be very small indeed, of the order of 0.003 reciprocal lattice unit – too small to draw on our usual scale of 10 cm to 1 reciprocal lattice unit. In electron diffraction work it is therefore customary to scale everything up by a factor $1/\lambda$, so that the reciprocal lattice parameters are of the form $a^* = 1/a$ Å^{-1} (in an appropriate orthogonal crystal system) instead of $a^* = \lambda/a$ dimensionless reciprocal lattice units. On this basis the sphere of reflection also has to be scaled up to $1/\lambda$ Å^{-1}, and is very large. Indeed it is so large that a small region of it round the origin can be approximated by a plane. The combination of the flatness of a portion of the sphere of reflection and the smearing of the reciprocal lattice points means that if the crystal is in an appropriate orientation a whole array of reciprocal lattice points (lying on a plane section of the reciprocal lattice through the origin) will be intersected simultaneously by the sphere of reflection (Fig. 16.2). It is therefore quite unnecessary to rotate or oscillate the crystal to obtain a diffraction pattern containing plenty of reflections; the specimen is in fact always kept stationary.

Yet a further simplification relative to X-ray diffraction arises from the short wavelength of electrons and the corresponding smallness of the Bragg angles. The spots produced on a flat film by the reflections are displaced from the origin by distances that are proportional to $\tan 2\theta$, and for such small angles $\tan 2\theta$ is proportional to $2 \sin \theta$. Thus distances on the film are directly proportional to distances in the reciprocal lattice. There is therefore no need to resort to elaborate methods of indexing the photographs; if we know the appropriate scale factor of the pattern and the orientation of the crystal then we can regard the electron diffraction pattern as a direct representation of a plane section of

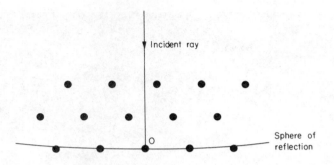

FIG. 16.2. The sphere of reflection, being of very large radius for electron diffraction, can intersect simultaneously a considerable number of reciprocal lattice points on a plane through the origin, O, and perpendicular to the page.

the reciprocal lattice. The spots can therefore be indexed just by counting them along the rows along which they lie.

The scale of the electron diffraction pattern depends on the magnification introduced by the projection lenses of the electron microscope, and of course any subsequent optical enlargement in printing the photograph. It can be calibrated by using a specimen coated with a thin film of aluminium or gold by evaporation *in vacuo*. The finely divided aluminium particles are in all orientations and so give a powder photograph consisting of concentric rings, as shown in Fig. 16.3. Since aluminium has a face-centred cubic structure with a unit cell edge of $a = 4.04$ Å, the first three powder lines are 111, 200 and 220 with d-spacings of 2.33, 2.02 and 1.43 Å. In the reciprocal lattice the 111, 200 and 220 points therefore lie at distances from the origin of 0.429, 0.495 and 0.699 Å^{-1}, and these are the radii of the circles formed by the powder lines. Comparison with the measured radii of the circles on the photograph (as in Fig. 16.3 itself) then gives the appropriate conversion factor to convert any measurement on the photograph to Å^{-1}.

The orientation of the crystal has to be found by trial and error. The specimen on its grid is mounted on a goniometer stage which can be tilted by controls outside the microscope, and the extent of such tilts can be measured. If the crystal is initially in an arbitrary orientation it may give very few diffraction spots, and it is tilted by trial and error until it gives a fairly dense array. The crystal is then known to be oriented so that the electron beam is perpendicular to a densely populated central section of the reciprocal lattice, which means that it is parallel to a simple zone axis of the crystal (e.g. [100], [001], [110], etc.). Since the probable nature of the crystal, and likely values for its unit cell dimensions are usually known in advance, it may then be possible to recognise the particular section of the reciprocal lattice that one has obtained. On the other hand, if the nature of the crystal is unknown the section may be designated an axial section, e.g. the x^*y^* section perpendicular to [001] and containing $hk0$ reflections. The indices can then be read off along the prominent layer lines as $0k0$, $1k0$, $2k0$, etc., and the reciprocal cell dimensions a^* and b^* measured. If the x^*- and y^*-axes are not orthogonal then the angle between them, γ^*, can also be measured directly. The crystal can then be tilted about one of the chosen axes until another dense array of diffraction spots is found; for example, if the crystal is tilted about the y^*-axis the beam will eventually lie along [101] and the rows of spots on

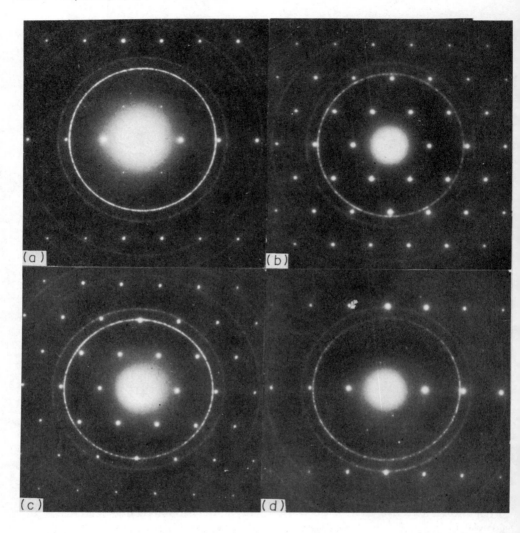

FIG. 16.3. Electron diffraction patterns of a single crystal of kaolinite (in four different orientations) coated with a thin film of aluminium. The spot pattern is from the kaolinite and the rings are the powder pattern of aluminium.

the photograph will be 0k0, 1k1, 2k2, etc. From the angles of tilt required to reach different reciprocal lattice sections in this way, and from measurements of the patterns, the remaining reciprocal lattice parameters c^*, α^*, β^* can all be derived.

There are, however, a number of ways in which caution has to be exercised in interpreting electron diffraction patterns. As we have seen, the strong interaction of electrons with matter means that the simple diffraction theory discussed in this book is a poor approximation to what happens in electron diffraction. One result of this is that the intensities of the reflections are not simply interpretable in terms of structure factors, and accordingly it is very rare for electron diffraction to be used for detailed crystal structure determination. Another result is that multiple reflections within a crystal can give rise to spots in the pattern in positions that correspond to reflections that are forbidden by the space group. Great care has to be taken in space group determination, therefore, to avoid errors arising in this way. Fortunately such "forbidden" reflections are usually weaker than the ordinary reflections, and it is also possible to distinguish them from true single reflections by a re-orientation of the crystal to a position in which they would only occur if they were permitted as single reflections. If they disappear in such circumstances then the space group absences can be correctly established.

Selected area electron diffraction patterns can be used to identify the nature of crystals observed in the electron microscope by reference to their cell dimensions and space groups. This is especially useful in the identification of very small inclusions of one mineral in another, such as frequently occurs as a result of exsolution of one phase from another. The method also reveals the relative crystallographic orientations of such contiguous crystalline phases.

These applications make use of the fact that electron diffraction permits the study of very small crystals that would be difficult or impossible to study otherwise than in the electron microscope, but another application makes a positive virtue of the fact that electron diffraction can only be applied to very small crystals or crystal fragments. This application is in investigating departures from regularity in crystals. Up to now we have assumed throughout this book that crystals consist of a perfect repetition of unit cells, but in fact a variety of imperfections may be present, some of which are discussed in Chapter 17. In X-ray diffraction the crystal is of the order of 0.1 mm in diameter, and so probably contains about 10^{15} unit cells. Any local defects in the regularity of repetition are therefore likely to have a negligible effect on the diffraction pattern, which effectively reveals the arrangement averaged over these 10^{15} cells. In electron diffraction, however, the volume of the crystal under observation may contain only a few thousand unit cells, and within this volume some particular departure from the average may be present to a substantial extent and have a significant effect on the diffraction pattern. A simple example would be if there were a tendency for alternate unit cells to differ in some way from one another. This would mean that the true unit cell would be twice as big, and so additional diffraction spots should appear half-way between the usual ones. However, if the tendency to alternation were only slight and were only maintained over short distances of a few unit cells, such extra spots might be undetectably weak and diffuse on an X-ray photograph, but on an electron diffraction pattern, from a very small volume of the crystal that happened to contain such an alternating region, they might be quite appreciable.

Although an extra set of spots such as that described above could easily be interpreted in terms of a doubling of the unit cell, the interpretation of the diffraction pattern arising from many other types of irregularity is much more difficult. Irregularities often lead to diffuse distributions of intensity in the diffraction pattern, and these can be very difficult to interpret unambiguously. Fortunately, the fact that electrons can be focused once again comes to our assistance. With electron microscopes of the highest resolution it is possible for the information contained in both the low-order Bragg reflections, and the diffuse intensity arising from irregularities, to be incorporated in the image forming process. In this way a high-resolution image showing the unit cells (Fig. 17.5), or even a certain amount of detail within the unit cells, may be obtained, and both the nature and locations of irregularities may be revealed. However, this direct imaging of the crystal lattice, and even some indication of the structure within the unit cells, does not supersede the interpretation of electron diffraction patterns. The production of such high-resolution micrographs requires very accurate orientation of the crystals, which can only be achieved by reference to their diffraction patterns; and spurious misleading detail in the micrographs can only be avoided by careful control of the focus of the electron microscope on the basis of complex diffraction calculations.

Problems

1. In Fig. 16.3(a) the kaolinite crystal was lying with its (001) plane perpendicular to the electron beam. The unit cell of kaolinite (although strictly triclinic) can be regarded for the purpose of this calculation as of approximately monoclinic shape with:

$$a = 5.15 \text{ Å}, \ b = 8.95 \text{ Å}, \ c = 7.35 \text{ Å}, \ \beta = 104.5°.$$

   Determine the scale of the photograph from the powder rings of aluminium (using the information given on p. 221) and index the strong reflections on the main layer lines (ignoring weak ones which may be due to multiple reflection within the crystal).
2. To obtain Fig. 16.3(b) the crystal was tilted through 14.5° about a line parallel to the horizontal (zero layer) line. Index the reflections on this photograph.
3. What Bravais lattice is suggested by the indices found in questions 1 and 2?
4. When the specimen is returned from position (b) through position (a) and tilted the other way, a succession of recognisable reciprocal lattice sections can be found of which Fig. 16.3(c) is the third. Find the angle of tilt from (a) to (c) and index the reflections on (c).
5. A further small tilt gave Fig. 16.3(d). What are the indices of the central spot on the upper layer line? To what crystallographic zone axis is the electron beam now parallel?

CHAPTER 17

Irregularities in Crystals

External form

CRYSTALLOGRAPHY started from the observation that crystals have smooth faces at definite angles to one another, whose directions are found to obey various symmetry relationships. From this it was deduced that crystals are built up from a regularly repeating pattern of atoms, and in the development of diffraction theory and space group theory this regular repetition is regarded as infinite in extent. Thus the bounding surfaces of the crystal, the very existence of the external form from which we started, come to be regarded as an "irregularity", a departure from regular repetition! For crystals big enough to see (bigger than 1 μm), the irregularity introduced by their boundaries is inappreciable even for X-ray diffraction methods; but below that size it introduces a broadening of the X-ray reflections that is inversely proportional to the crystal dimensions, as discussed in Chapter 9. A brief discussion of some of the factors that control external form will be found in Chapter 18.

In addition to limiting the extent of the ideally infinite repeating pattern of a crystal structure, the faces of a crystal frequently reveal more serious signs of irregularity. It has already been pointed out in Chapter 7 that the faces of a crystal do not all look alike. Some faces are indeed optically flat as would be expected if they are atomically smooth, or if they consist of steps with unit cell dimensions. However, some faces are matt on account of small-scale pits or hummocks; others are composed of a mosaic of small optically bright areas that are misoriented from one another at small angles, and many are striated in a variety of ways. It is possible that some departures from smooth faces could occur even on a crystal that was internally perfect, as a result of fluctuations in the temperature or composition of the fluid from which it grew, and this could account for some kinds of striations. However, most surface imperfections probably arise from internal imperfections of various kinds. Segregations of impurities can lead to the growth of matt and humocky faces, which may also arise from incipient solution taking place around impurities or around dislocations. The latter may also give rise to growth steps (and hence to striations) or to a mosaic structure with low angle boundaries. Striations can also arise from polysynthetic twinning. Structural imperfections of these various kinds are discussed in this chapter.

Cleavage and parting

Many crystals, but by no means all, exhibit the property of *cleavage*; that is to say that when they are appropriately stressed they split along crystallographic planes of particular indices. This happens when there is significantly weaker bonding in the structure across

226   Crystallography for Earth Science Students

certain sets of lattice planes than in other directions. The cleavage faces thus produced may be very perfectly smooth and reflecting, though they may also show a mosaic of areas that are misoriented from one another at small angles; as in the case of natural faces, this arises from misorientations in the structure of the crystal which are discussed below.

The cleavage may reveal irregularities in a crystal, but it is in itself an expression of the regularity of the structure. There is, however, another related phenomenon known as *parting* that results entirely from irregularities in the structure. Parting is superficially similar to cleavage, in that it is a preferential fracture parallel to a particular lattice plane. But, whereas cleavage to give an (*hkl*) face can occur on any lattice plane of the set *hkl*, a parting that gives a face (*hkl*) occurs only at specific, though apparently random, positions in the crystal. It arises not from planes of weakness in the regular structure but from specific planes of weakness arising from irregularities in the structure. These may be due to segregation of impurities, for example, or to such irregularities as twin boundaries and stacking faults which are discussed below.

Twinned crystals

A more serious irregularity makes itself evident in the external form of some *twinned* crystals such as those shown in Fig. 17.1. Such twins contain two or more portions which

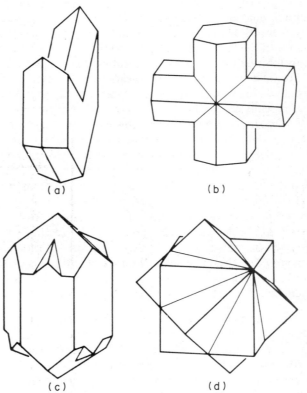

FIG. 17.1 (a) Contact twin of gypsum. (b) Cross twin of staurolite. (c) "Dauphiné" twin of quartz. (d) Interpenetrant twin of fluorite.

are physically continuous with one another but which are in different crystallographic orientations. This may also appear to be true of the very common polycrystalline aggregates like that shown in Fig. 17.2. These are produced when a group of crystals, initially growing independently and in unrelated orientations, meet and adhere strongly to one another. But twinned crystals are distinct from polycrystalline aggregates in that their components are in orientations that are symmetrically related to one another, and not random. In the crystals shown in Fig. 17.1 it is evident from the orientation of the faces (which often, though not necessarily, form re-entrant angles on twinned crystals) that there must be an irregularity in the lattice repetition where the components of the twin meet.

FIG. 17.2. A group of intergrown crystals (quartz).

Twinning has been studied for almost as long as crystallography itself, but it is a part of the subject which is still poorly understood. There has even been controversy in recent years as to the precise definition of twinning and therefore the scope of the phenomena to be included. Most descriptions of the subject tend to classify twins mainly (and often in great detail) in terms of the symmetry relationship between the members of the twin, and to ignore the less well-understood question of why and how twins form; as a result they tend to leave the reader with a sense of mystification.

It appears probable that the elements of a twin may be related by any crystallographic symmetry element (i.e. 2-, 3-, 4- or 6-axes of rotation, $\bar{3}, \bar{4}$ and $\bar{6}$-axes of rotation–inversion, a mirror plane or a $\bar{1}$ centre of symmetry), but in the overwhelming majority of cases the relationship is by a mirror plane or a 2-fold axis. The only fundamental restriction is that the relationship cannot be by a symmetry element that is possessed by the crystal itself in that orientation, because then the symmetry operation would transform the crystal into itself and not into its twin. Because the combination of a $\bar{1}$ (centre of symmetry) with a 2-axis gives rise to a perpendicular mirror plane and vice versa, it follows that in a crystal belonging to any centro-symmetric point group, members of a twin that are related by a mirror plane are also related by a 2-fold axis and vice versa.

The symmetry relationship between the members of a twin is commonly described as a twin law. To specify this twin law one quotes the indices of the crystallographic plane (hkl) parallel to the (mirror) plane, or the zone axis $[uvw]$ parallel to the symmetry axis, that relates the components of the twin to one another. If the symmetry axis is not 2-fold it is also necessary to specify its order. Such a plane or axis is described as the *twin plane* or *twin axis*. In some twins (Fig. 17.1(a)) the members are in contact with one another across a crystallographic plane (often, but not always, the twin plane) and this plane is then called the *composition plane*, but not all twins possess a composition plane (e.g. Fig. 17.1(c)) for reasons that we shall see.

The most obvious kinds of twins are those in which the twinning is directly evident from the facial development, like those illustrated in Fig. 17.1. However, the fundamental criterion of twinning is the presence of domains whose crystal lattices are differently, but symmetrically, oriented and this may not always be visible from the external form. In such cases the twinning may sometimes be detectable from related patterns of striations on certain faces, from the observation of different orientations of the optic axes (or other directional physical properties) of anisotropic crystals, or, in every case, by diffraction methods. The phenomena of twinning may best be understood in terms of the following classification of twins into four categories depending on the processes involved in their formation. This cuts across classifications based either on symmetry relationships or external appearances.

(i) *Simple growth twinning*

Crystals normally grow by the deposition of successive layers of atoms on a growing face, and the deposited atoms take up the lowest energy positions available to them. If this process is followed strictly then a single crystal is formed. However, there may be an alternative set of positions in which the depositing atoms would have only a slightly higher energy, and there is then a finite probability that they will take up these positions. Figure 17.3 shows a very simplified schematic example. The layer of atoms B' has the same distance relationship to its nearest neighbours at A' as does the layer B to layer A, and the fact that distances from its more distant neighbours in the layer below A' are "wrong" will have only a minor effect on its energy. There may therefore be a finite probability of laying down a layer like B'. When layer C' is laid down, the lowest energy position will be achieved if it is related to *both* B' and A' in the way that C is related to B and A. It therefore goes into a position which continues a twin domain in an orientation reflected in a horizontal twin

FIG. 17.3. Contact twin relationship between two crystalline arrays of atoms.

plane. Such a process obviously gives rise to a contact twin. If the energy difference between the alternative position of B' on A' from B on A is appreciable then the twinning process may only occur occasionally, but if it is very small the twinning process will occur many times at random intervals within every crystal, to give what is called *polysynthetic twinning*. If the twinning happens only once or a few times during the growth of the crystal then it may be very evident in the morphology, but if it happens very frequently the faces in twinned orientation may be so narrow that they appear only as striations, or are totally invisible.

(ii) *Shear twinning*

Calcite is the best-known example of a mineral that twins by mechanical shear. If a cleavage rhomb is held with an obtuse edge horizontal and a sharp knife is pressed at right angles to this edge, a portion of the crystal can be sheared into a twin position as shown in Fig. 17.4. The twinned portion is related to the original crystal by reflection across the horizontal plane $(01\bar{1}2)$, which is both the twin plane and the composition plane. Any other way of applying a shear stress in the same direction could lead to twinning. The process depends on the same structural features as simple growth twinning, namely that there is an alternative available relationship between adjacent layers in the structure; but it also requires an additional criterion to be satisfied, namely that there is a low-energy path for a layer to shift from one position to the other. This permits the shear to move successive layers one at a time, into the twinned position, without disrupting the crystal.

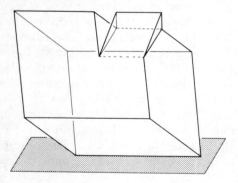

FIG. 17.4. Shear twinning of calcite. Twin (and composition) plane shown by broken lines.

(iii) *Transformation twinning*

There are many materials that have different stable crystalline forms (polymorphs) at high and low temperatures, the high-temperature form being of higher symmetry. If the difference in structure of the two forms is slight they may transform easily and reversibly at an *inversion temperature*. Then, on cooling the high-temperature form through the inversion temperature, domains of the low temperature form may start to grow in different orientations in different parts of the crystal, these different orientations being related by the symmetry of the high-temperature form. The domains are therefore in twinned orientations relative to one another, but they meet along more or less haphazard boundaries depending on the accidents of where they started to form. There may also be many domains or few, depending on whether or not the transformation started simultaneously at many different centres. Obviously such a transformation will not of itself give rise to new faces on the crystal, which will therefore initially have the same morphology as that of the higher symmetry polymorph, but if it continues to grow after transformation then it may develop new faces growing in different directions where different domains happen to emerge at the surface. It can therefore give rise to interpenetrant twins as in Fig. 17.1(c). A good example of this type of twinning is the so-called Dauphiné twinning of quartz. Above 573°C quartz has the point group 622, whereas below this transition temperature it has the point group 32. When low quartz is formed from high quartz at the transition, its x-, y-, w-axes may coincide with a set of axes of the parent high quartz, or may lie along the directions at 60° to those that are symmetrically equivalent to them in high quartz. Domains formed in these two orientations are therefore related by a rotation of 180° about the z-axis. Similar twinning processes (including, indeed, this Dauphiné twinning of quartz) can also be induced by mechanical stress of certain kinds.

(iv) *Nucleation twinning*

There remain very many cases of twinning, including many of the most spectacular ones, which cannot be explained by any of the preceding mechanisms. If a substance habitually forms contact twins like Fig. 17.1(a) this cannot be due to simple growth twinning (or shear twinning), for if the twinning process were sufficiently probable to occur in so many crystals it would be bound to occur several times, or many times, in some of them. Again, interpenetrant twins of fluorite have their members related by a 180° rotation about one of the [111] cube diagonals (Fig. 17.1(d)), but they are never twinned again about any of their other symmetrically equivalent cube diagonals, which shows that the twinning cannot have arisen as a random process during growth. The twin pair taken together has a 6-fold axis along the twin axis, but there is no known structure of fluorite with hexagonal symmetry that could have given rise to transformation twinning. The solution to the problem posed by twins of these kinds seems to reside in the nucleus from which the crystal grew. This nucleus must have differed in structure from the crystals that grew from it; thus the nucleus of Fig. 17.1(a) must have had a plane of symmetry, and that of Fig. 17.1(b) must have had an approximate* 4-fold axis, and that of Fig. 17.1(d) a 6-fold axis.

---

* This symmetry is not exact, as the arms of the cross are not exactly perpendicular.

Irregularities in Crystals 231

It would be exceedingly difficult to identify the exact nature of the nucleus from which a crystal grew. It might contain one or more impurity atoms that influenced its structure. On the other hand, it is quite possible that the most stable configuration of a small group of atoms may be such that it cannot repeat regularly into a crystal structure. In such a case the only way that it can grow is for crystalline regions (of different but related structure) to grow on to it. If the regions that grow from different parts of the nucleus are in different but symmetrically related orientations, then a twin will result. This type of twinning may lead to a contact twin (Fig. 17.1(a)) if the nucleus is essentially lamellar, or to interpenetrant twins (b or d) if it has axial symmetry. Such interpenetrant twins may be united across composition planes as in (b), or not as in (d), depending on the relationship of the twin axis to the symmetry of the crystals, but even when composition planes are possible as in (b) their planarity depends on equal rates of growth of the regions at each side of them. Composition planes of this sort are in a sense accidental compared with those in simple growth twins and shear twins, in which it is the fit of the structure across the composition plane that is the basic cause of the twinning.

The irregularities introduced by twinning are confined to about one atomic diameter to each side of the surface of contact between the twinned individuals (whether this is a composition plane or not). They therefore occupy an exceedingly small fraction of the total volume of the twinned crystal, and are accordingly very difficult to investigate in detail by diffraction or otherwise. In general their nature can only be conjectured from what is known of the regular crystal structure, combined with the symmetry properties of

FIG. 17.5. High resolution electron micrograph of grunerite resolving the lattice structure. Very frequent polysynthetic twining is revealed with a c-glide reflection on 100 planes (vertical lines in the photograph).

the twins involved. Occasionally, however, it is possible to obtain some direct evidence by the technique of high-resolution electron microscopy described in Chapter 16. Figure 17.5 is such a micrograph of highly twinned grunerite (a chain silicate), in which the structure is resolved to the extent of revealing two fuzzy blobs per unit cell. The twinning involves reflection in vertical lines on the micrograph, but it may be seen that the twin operation is not in fact a mirror reflection but a glide reflection, the blobs on opposite sides of the composition planes being displaced relatively by half a unit cell in the vertical direction. This glide relationship can in fact be readily explained in terms of the details of the crystal structure.

Polytypes and stacking disorder

We have seen that in some crystal structures there may be alternative sets of sites on a growing face which offer only slight differences in their bonding energy to the next layer of atoms, and this can lead to simple growth twinning. This situation arises in extreme form in layer silicates like the micas and clay minerals. In the structures of these minerals the atoms are strongly bonded together into two-dimensional sheets 5–10 atoms thick, but there is only very weak bonding from one such sheet to the next. Not only is the interlayer bonding weak, but the surfaces of the sheets are rather more symmetrical than the structure as a whole. There are therefore often several alternative ways of stacking one layer on another which lead to structures of very similar stability. If the same choice of stacking position is made at each layer a regular crystal structure results, but there are a number of such possible regular structures depending on this choice. Furthermore, different choices may be made in a regular sequence. Thus with two choices denoted A and B we may obtain not only the obvious regular stacking sequences, $AAAAA$ and $BBBBB$, but also $ABABAB$ or $AABAABAAB$. Such alternative stackings of the same kind of layers are related to one another more closely than polymorphs (which may have totally different kinds of structures), and they are called *polytypes*. The diffraction patterns of polytypes of the same material are perfectly normal, but their unit cells are closely related geometrically.

Such polytypes do not in themselves involve any kind of irregularity; indeed they sometimes involve a surprisingly long-range regularity in the sequence of stacking choices. However, materials with layer structures of this type, which provide alternative stacking arrangements of similar energies, are naturally very liable to deviate from regularity, either occasionally to give an effect rather like a twin, or else so frequently that there is no regularity in the stacking pattern – apart from the fact that each layer is at the same perpendicular distance from the next—as shown schematically in Fig. 17.6. Such irregularity has a profound effect on the diffraction pattern. If the layers of the structure are taken to be parallel to (001), then reflections of the type 00l will be quite normal, because of the regularity of stacking along [001]; indeed all 00l planes will pass through the lattice points in the usual way. However, if the lateral displacements of the layers are totally irregular it is impossible to define lattice planes with more general indices (such as $hk0$ or hkl). Therefore, apart from the normal 00l reflections, the only other diffraction effects are those that arise from the two-dimensional regularity within each layer, considered independently. Such two-dimensionally repeating structures give diffraction patterns, but the diffracted rays are not confined to specific Bragg angles; instead of sharp lines on a powder pattern they give rise to diffuse bands which cut off sharply on the low-

FIG. 17.6. Random stacking (two-dimensional analogue).

angle side and die away gradually on the high-angle side (Fig. 17.7). They can be characterised by two indices hk, and the low angle cut off is close to the position that would be expected for a proper $hk0$ reflection from a corresponding three-dimensionally regular structure. A material whose structure consists of equally spaced parallel layers that are subject to random displacements and rotations in their own planes is said to be *turbostratic*, and its powder diffraction pattern consists of a single set of sharp $00l$ lines

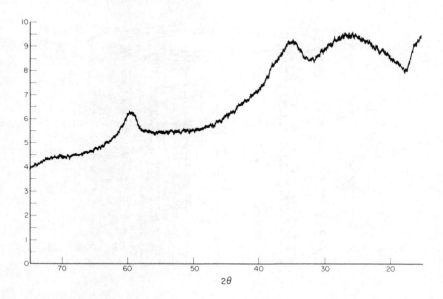

FIG. 17.7. X-ray diffractometer trace of a material (magnesium trisilicate) that consists of unstacked silicate layers.

(successive orders of the basal spacing) and a set of asymmetric *hk* bands. Examples of such a structure are provided by various kinds of "amorphous" carbon. Many so-called amorphous materials have some measure of order in their structure, even though it falls short of a fully crystalline order. A more extreme case is "amorphous" magnesium trisilicate, which gives a powder pattern (Fig. 17.7) consisting only of a series of asymmetric diffuse bands, indicating that the atoms are arranged into two-dimensional layers, resembling those in a layer silicate, but these layers are not even stacked parallel to one another, with the result that there are no sharp 00*l* reflections.

Possibilities also exist for degrees of order intermediate between the turbostratic and the fully crystalline. Parallel layers stacked at equal intervals may suffer relative displacements in their own planes in one direction only (say the *y*-axis) and be undisplaced parallel to the *x*-axis. In such a case any set of crystallographic planes parallel to the *y*-axis will pass through every lattice point and will give rise to perfectly normal X-ray reflections, so that the diffraction pattern will contain sharp *h*0*l* lines (rather than just the 00*l* lines of a turbostratic structure) together with diffuse asymmetric *hk* bands. An example of such a structure is that of the serpentine mineral chrysotile, the main kind of commercial asbestos. This is a layer silicate in which the layers have an inherent tendency to curl. Because they wrap round into cylindrical rolls (Fig. 17.8), each roll constituting an asbestos fibre, successive layers can be ordered with respect to one another in the direction parallel to the cylinder axis. However, in the circumferential direction each layer is bent to a different radius and is unable to keep in step with its neighbours, and so complete disorder in the circumferential direction results. Although the diffraction pattern of chrysotile is further affected by the fact that the silicate layers are curved and not flat, its main features are those described – sharp *h*0*l* reflections and diffuse *hk* bands.

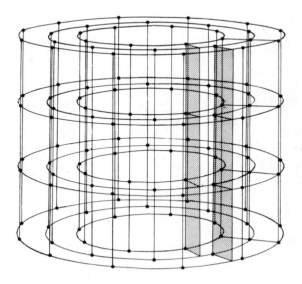

FIG. 17.8. Circumferential disorder of successive layers (shown as cylindrically curved two-dimensional lattices) in chrysotile asbestos. A projection of the structure on to a radial plane (shaded) is coherent.

Many clay minerals contain a further degree of order, illustrated in Fig. 17.9, that still falls short of full crystalline order. Here the displacements of the layers parallel to the y-axis are limited to steps of $\pm b/3$ instead of being completely random. As a result, sets of planes having indices hkl, with k a multiple of 3, pass through every lattice point and give rise to normal X-ray reflections. Thus the diffraction pattern will contain sharp reflections of type $h0l$, $h3l$, $h6l$, etc., but diffuse bands of type $h1$, $h2$, $h4$, $h5$, etc. If the stepwise disorder were in steps of b/r then the sharp reflections would be those in which k was a multiple of r.

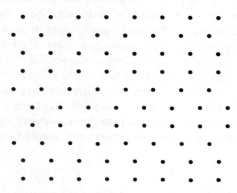

FIG. 17.9. Stepwise displacement of layers.

Irregularities of composition

In the treatment of irregularities in crystals up to this point we have assumed that they involve only irregularities in the positions of units that are chemically identical throughout the structure. However, this is by no means always true. Indeed, silicate minerals are usually solid solutions containing non-integral numbers of atoms of various elements in the formula unit, which means that the unit cells of the crystal structure cannot all be of the same composition.

The effect of this on the diffraction pattern has already been discussed in Chapter 15. The structure can be regarded as the sum of an averaged structure and a difference structure, and the diffracted amplitude is the sum of the diffraction effects from these two components. Diffraction from the averaged structure is confined to the sharp Bragg reflections, where it is very strong, whereas that from the difference structure occurs at all angles but is weak everywhere.

However, this conclusion is based on the supposition that the deviations from the average structure are entirely random. This is not always the case; sometimes there is some degree of order. For example, there may be a tendency for alternation in the composition of adjacent unit cells, though with the regularity of alternation only extending over small domains. This leads to a difference pattern in which there are small domains of regular structure with a larger unit cell, and these will give a crystalline type of diffraction pattern at appropriate Bragg angles, but because of the small size of the domains these reflections will be very much broadened for the same reason as the reflections from very small crystals (p. 143). The result is that diffuse reflections are observed between the sharp reflections, although these diffuse reflections are much less diffuse than the scattering from a random difference structure – that is why they can be readily observed. Thus the diffraction patterns of perfectly ordered structures and of perfectly random solid solutions both consist of sharp reflections only, whereas partially ordered solid solutions give more complex patterns involving additional diffuse reflections. Such phenomena are common in the feldspars.

The mixed-layer clay minerals provide another example of a class of structures in which the unit cells vary in size and composition, but to a much greater extent than that which we have been considering. In these materials there may be a (more or less) random interstratification of silicate layers of different thicknesses such as 10 Å and 14 Å, or 10 Å and 17 Å. Basal 00l reflections from such structures tend to occur at positions corresponding to the weighted mean spacing, corresponding to the averaged structure discussed above, but because of the large discrepancy between the constituent layer thicknesses the diffraction effects from the difference structures are by no means negligible. As a result the reflections often spread diffusely from positions appropriate to the average structure to positions appropriate to the constituent layer thicknesses, and the intensity maxima do not always occur at regular increments of $\sin \theta$.

Crystal defects

Although crystals containing the various kinds of irregularity that have been discussed in this chapter are obviously defective in various ways, the term "crystal defect" has come to have specific reference to *point defects* and *dislocations*.

Point defects need no further discussion here. This term includes *vacancies*, where an occasional atom is omitted from its appropriate site in the structure, and *interstitials* where an additional, uncovenanted, atom is inserted. Such defects lead to the same sort of disturbance of the structure as the random substitutions already discussed, and like them do not lead to any appreciable effect on the diffraction pattern.

Edge dislocations correspond to the termination of a layer of the crystal structure along a line within the crystal, as shown in a cross-section perpendicular to such an edge dislocation in Fig. 17.10. A series of such edge dislocations may give rise to a *low-angle*

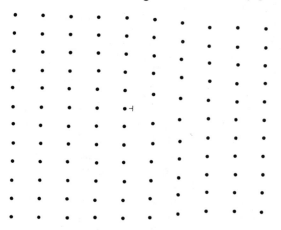

FIG. 17.10. An edge dislocation viewed end-on.

Irregularities in Crystals   237

boundary between domains of the crystal, these domains being slightly misoriented with respect to one another (Fig. 17.11). *Screw dislocations* involve a helical twist in a structural layer so that a continuation of a single layer is able to grow over itself and build up into a whole crystal (Fig. 17.12). Both types of dislocation therefore lead to misorientations of

FIG. 17.11. A low-angle boundary between two domains, produced by a series of edge dislocations.

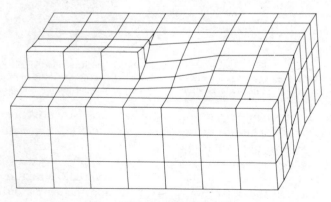

FIG. 17.12. A screw dislocation.

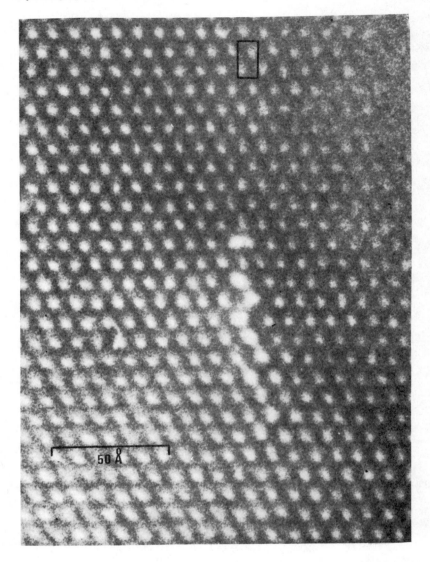

FIG. 17.13. An edge dislocation seen in a high-resolution electron micrograph of grunerite viewed down the z-axis. The unit cell outlined forms part of a vertical row that terminates in the neighbourhood of the large white spots in the middle of the photograph.

parts of the lattice, and so limit the regions of the crystal over which coherent diffraction takes place. Such misorientations of crystal domains, constituting a so-called mosaic structure, do not lead to any departures from regular diffraction patterns, since the domains will all reflect rays at the same Bragg angle, but consecutively rather than simultaneously as the crystal is rotated. Indeed the mosaic structure improves the correspondence between the diffraction phenomena and the simple diffraction theory described in this book because it reduces the extent to which underlying parts of a crystal

FIG. 17.14. Spiral growth steps on a crystal face due to the emergence of a screw dislocation (from Gritzayenko, G. C. et al., *Methods of Electron Microscopy of Minerals*, Moscow, 1969). Scale bar 50 μm.

are shielded from the incident beam as a result of reflection in overlying layers. As pointed out on p. 217, we have implicitly assumed this shielding to be negligible for X-rays, so that a circumstance that reduces it improves the applicability of the theory.

Edge dislocations can be seen by high-resolution electron microscopy of a thin section of the crystal viewed parallel to the dislocation (Fig. 17.13). Screw dislocations can be seen by optical microscopy of crystal faces which they intersect, because they give rise to low spiral steps on the face (Fig. 17.14).

CHAPTER 18

Morphology Revisited

This book started from a consideration of the external morphology of crystals, which is their most obvious characteristic and constituted the historical starting-point of the science of crystallography. The concepts that have been developed all stem from the existence of zones of faces, and provide a logical general explanation of them. Up to this point we have, however, said very little that explains the particular morphology of any particular crystal. It is in principle possible to explain the disposition of atoms in a crystal structure in terms of a minimum energy configuration, if the forces between the atoms can be calculated. The disposition of atoms that satisfies this criterion then necessarily conforms to a particular space-group, and the morphology of the crystal therefore conforms to the crystal system and crystal class to which that space-group belongs. There are, however, many alternative morphologies open to a crystal that belongs to any particular class, and nothing that has been said so far provides any basis for expecting a crystal to be bounded by one set of forms rather than another. Amongst cubic minerals, for example, one may wonder why fluorite favours the cube form $\{100\}$, magnetite the octahedral form $\{111\}$, and garnet the dodecahedral form $\{110\}$. Again, amongst tetragonal crystals one may wonder why the tetragonal prism forms may dominate over the pyramidal or basal forms to give slender acicular crystals in rutile, or on the other hand the basal $\{001\}$ form may be prominent and give stubby (or sometimes even tabular) crystals as in apophyllite. Moreover, any answer to such questions must take account of the fact that a given mineral from different localities may show quite different forms; thus fluorite may form octahedra instead of cubes, and combinations of the cube with forms such as $\{210\}$ or even with a general form $\{421\}$; while calcite occurs in a great variety of different habits.

It is not possible to give general answers to all these questions, nor even complete answers for any particular mineral, but some attempt must clearly be made to indicate the directions in which answers may be sought. The first step is to consider that for any particular crystal structure there will be some particular morphology corresponding to the lowest energy configuration. Within the crystal the configuration of the atoms will already correspond to an energy minimum due to their interactions, but on a crystal face this arrangement is necessarily disturbed and the atoms are incompletely bonded. Thus each face will have associated with it a surface energy proportional to its area, and the equilibrium morphology will be that which minimises the total surface energy for a given volume; it may involve small faces of relatively high energy if the presence of these permits a larger offsetting reduction in the area of faces of relatively low energy.

In very general terms the surface energy of a form $\{hkl\}$ may be expected to be lower the greater the number of lattice points per unit area on a plane *hkl*. This is because a unit of pattern on such a face of high *reticular density* will in general be more thoroughly bonded to its neighbours than one on a face of lower reticular density. This is the basis of the Law of Bravais that the importance of a face is proportional to its reticular density, and it is a sufficiently good approximation that the statistical importance of different forms averaged over many crystals from different types of occurrence is usually in descending order of reticular density. Since the reticular density on different planes depends on the Bravais lattice, such morphological statistics would frequently permit the assignment of Bravais lattice in absence of X-ray information. An alternative, and equivalent, statement of the law of Bravais is that the importance of a form $\{hkl\}$ is proportional to d_{hkl}. There is therefore a distinct tendency for the habit of crystals to bear an inverse relationship to the shape of the unit cell; those with elongated cells form tabular crystals with their thickness in the direction of the greatest cell edge, and those with flattened cells form acicular crystals elongated parallel to the smallest cell edge.

There are, however, many causes of departure from the Law of Bravais. It is an excessive simplification to regard the "motif" of the repeating pattern of the crystal structure as a physical unit that is bonded to its neighbours. A further degree of approximation is attained in the modification of the Law of Bravais called the Donnay–Harker Law. This regards the effective value of d_{hkl} controlling the importance of the form $\{hkl\}$ as being reduced by any screw axis or glide translation perpendicular to *hkl*, the factor of reduction, n, being the same as the first non-extinguished order of X-ray reflection from *hkl*, because the spacing d_{hkl}/n is the spacing between layers of sub-units of the pattern that are related by the symmetry. Ultimately, however, it must be admitted that the surface energies of different forms cannot be accounted for in terms of units or sub-units of pattern in a generalised sense, but must be discussed in terms of the actual interatomic bonding and the deficiency of bonding, and the nature of the atoms exposed, on a face. For example, a (111) face of halite would consist either of all Na ions or of all Cl ions and so would be highly charged and of high energy, whereas a (100) face contains equal numbers of ions of opposite charge. By contrast a (111) face of sphalerite lies close to both Zn and S atoms in the structure. It is therefore understandable that these two minerals, both with an F-centred Bravais lattice, favour different habits – cubes for halite and tetrahedra for sphalerite. Every substance therefore needs individual consideration in terms of its structure, and indeed of such adjustments to the structure as may occur in the immediate subsurface layer (and these are still largely unknown). The relative surface energies of the different forms may also be modified by the nature of the surrounding fluid.

Even if all these factors could be taken into account, however, only the equilibrium morphology would be defined, and in fact it must be expected that crystals will only rarely attain such an equilibrium state. More usually their morphology will be affected by the relative rates of addition of material to the various faces. We have already seen in Chapter 7 how such an effect can lead to differences in the morphology of crystals belonging to holosymmetric and lower symmetry classes. This result is obviously a contravention of the Law of Bravais, since all the faces of any given holosymmetric form still have the same reticular density when they split into separate forms in a crystal class of lower symmetry. Thus the mechanism invoked to explain low symmetry morphology, namely differing impediments to growth on different forms arising from adsorption of foreign atoms, ions or molecules, is a mechanism of general validity which is undoubtedly responsible for much of the variability in the morphology of a given mineral formed in different environments. Another somewhat related mechanism is also sensitive to the chemical composition of the fluid from which a crystal grows. The deposition of an additional layer of unit cells on a crystal face may involve several successive processes of depositing different constituents of the structure—for example, alternating layers of cations and anions. These individual processes may be of varying degrees of difficulty (both energetically and sterically) and the most difficult will be the rate-controlling step. Changes in the composition of the fluid may change the relative availability of the different constituents of the crystal for deposition on the face, and so may change the nature of the step that is rate controlling. Since such effects will be different on different faces, the chemistry of the fluid may change the morphology. It is known, for example, that halite forms cubes at low pH and octahedra at high pH.

In Chapter 5 we invoked physical barriers to growth to explain the existence of ill-formed crystals. Such physical effects may be more subtle however. If the supplying of "nutrient" for crystal growth (by diffusion or convection) is directional along a direction parallel to a zone axis it may promote the growth of long acicular or fibrous crystals. Rapid crystal growth into more concentrated regions of the fluid may effectively deprive other regions of the crystal of access to the nutrient, and this leads to the formation of oriented intergrowths, hopper crystals and dendritic growths (Fig. 18.1). These are all single crystals, in that the structure is continuous and in a single orientation throughout. The facets with particular indices are all parallel, although not contiguous, being separated from one another by re-entrant angles. Such angles are particularly favourable to further growth, since a deposition there is more securely attached to two faces instead of one; re-entrant angles can therefore only be preserved if the projecting parts of the crystal have grown so fast as to shield them from access by nutrient. Such growths are therefore promoted by high degrees of supersaturation of a solution (or supercooling of a melt) from which the crystals are growing.

In addition to the primary structure of the crystal and the composition of the solution or melt, screw dislocations can also have an important role in the growth of a crystal. This is because a re-entrant angle, as we have seen, provides a much more secure and favourable

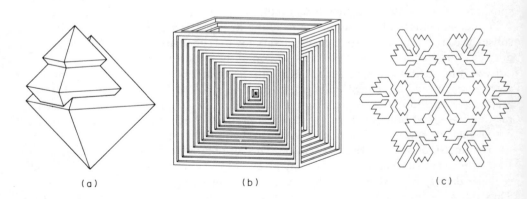

FIG. 18.1. (a) Oriented intergrowths, (b) hopper crystals and (c) dendritic growths. (From *An Introduction to Crystallography* by F. C. Phillips, 4th ed., Longmans, 1971.)

location than a flat surface for addition of an atom or group of atoms to a growing crystal. This is true even if the re-entrant angle is simply at the base of a step that is only a single unit cell in height. Such a step occurs wherever a screw dislocation emerges at a crystal face (Fig. 17.12). As material deposits at the step and moves it forward a new layer is deposited but the step never disappears; when a complete new layer has spread over the face the step has simply rotated round the dislocation and is back in the same position. However, in general, the rate of advance of the step varies with its distance from the dislocation itself, and so the step winds itself round into a spiral which may be detected by microscopy on many crystal faces (Fig. 17.14). The concentration of screw dislocations on crystal faces is itself dependent on the structure, and also on the conditions in which the crystal nucleated and the mechanical strains to which it has been subjected.

The multifarious factors that control the morphology of crystals thus have a great influence on their variety and beauty. And although logical arguments have led from the basic facts of crystal morphology to an exceedingly elegant and useful science of crystallography, this science is still far from being able to predict in detail the shape of any individual crystal that may be formed in arbitrary circumstances. Indeed, alongside the science, there is still a natural history of crystal and mineral forms. The aesthetic attraction of crystals remains, however great our knowledge of their intimate structure.

Glossary of Terms Used in Crystallography

Atomic scattering factor (f): The amplitude of a wave scattered by an atom in a particular direction relative to the amplitude of a wave similarly scattered by a classical electron.

Axes of a crystal: Axes parallel to the edges of the unit cell.

Axis of rotation symmetry: A line about which an operation of rotation through $360°/n$ is applied. In crystallography n can only have the values (1), 2, 3, 4 or 6.

Axis of rotation–inversion: The symmetry element corresponding to the operation that *both* rotates through $360°/n$ *and* then inverts through the origin.

Bevel: If the edge between two faces of a form is replaced by two small faces, each of which makes an equal angle with its adjacent larger face, then the pair of small faces is said to bevel the edge.

Bragg angle: The angle between an incident ray and a lattice plane for which the Bragg equation is satisfied.

Bragg equation: The equation $\lambda = 2d \sin \theta$ which specifies the relation between wavelength (λ), interplanar spacing (d) and angle of incidence (θ) of a ray on a set of lattice planes, such that reflections from all planes of the set are in phase with one another.

Centre of symmetry: The point symmetry element about which the operation is applied that inverts every point x, y, z to $-x, -y, -z$.

Centred cell: A cell of a lattice with lattice points at either the body centre of the cell, or at the centres of two opposite faces or at the centres of all the faces.

Circle of reflection: A circle in which the sphere of reflection is intersected by a plane of the reciprocal lattice.

Constancy of angles: Faces that recognisably correspond to one another on different crystals of the same substance make equal interfacial angles with each other.

Crystal class: Two crystals belong to the same class if they have all the same symmetry elements.

Crystal system: Two crystals belong to the same system if they have cells whose shapes possess the same symmetry.

Diad: A 2-fold rotation axis, *or* a 2-fold rotation–inversion axis (which is equivalent to a perpendicular mirror plane).

Direction cosines: The cosines of the three angles made by a line with three orthogonal coordinate axes. Conventionally denoted l, m, n, they define the direction of the line.

Edge: The line of intersection of two faces.

Enantiomorphous: Having no element of symmetry (centre or mirror) that changes "hand". As a result an enantiomorphous object can occur in two configurations, either right-handed or left-handed.

Face: A natural flat surface on a crystal, that develops as a result of the internal structure.

Form: A set of faces related to each other by the symmetry of the crystal.

Form indices: The Miller indices of a face enclosed in curly brackets $\{hkl\}$, interpreted to imply all the faces of the form to which the face belongs. The indices of the other faces in the form are obtained by permuting these indices (and their signs) to the extent that is demanded by the symmetry.

Glide plane: A symmetry element in a repeating pattern whose operation is a combination of reflection in the plane and a shift of the reflected image parallel to the plane by half of a lattice translation.

Habit: The overall shape of the crystal, described either generally or in terms of a dominant form or forms.

Hexad: A 6-fold rotation axis or rotation–inversion axis.

Holosymmetric: Having the maximum symmetry possible in a particular crystal system.

Ill-formed crystal: A crystal on which different faces of the same form are of unequal size, so that the shape obscures the true symmetry.

Indices of a face (Miller indices): A set of three whole numbers (hkl) such that the intercepts of a face on axes x, y, z parallel to the edges a, b, c of the unit cell are in the ratio

$$\frac{a}{h} : \frac{b}{k} : \frac{c}{l}.$$

Interfacial angle: The angle between two faces, defined as the angle between lines perpendicular to them.

Interplanar spacing: The perpendicular distance between adjacent planes in a set of lattice planes.

Glossary of Terms Used in Crystallography

Lattice: The imaginary array of points in space whose repetition corresponds to the repetition of the motif of a repeating pattern.

Lattice planes (set of): A set of parallel, equally spaced, planes that pass through every point of a lattice.

Mirror plane: The symmetry element corresponding to the operation that converts an object into its mirror image.

Monochromatic X-rays: X-rays having (approximately) a single wavelength.

Pinacoid: A pair of parallel faces related by symmetry to one another but not to any other face.

Point group: A self-consistent combination of symmetry operations around a point, and therefore appropriate to the external symmetry of a crystal. Each crystal class corresponds to a point group.

Polar: A crystal having at least one axis whose two ends are not related by symmetry. This implies lack of a centre of symmetry.

Primitive cell: A cell of a lattice with lattice points at the corners of the cell only.

Primitive of the stereographic projection: The projection of the equator of the sphere on the stereographic projection.

Prism: A set of faces parallel to an axis that are related by symmetry to one another but not to any other face.

Reciprocal lattice: An imaginary lattice whose points are (relative to an arbitrary origin) at the ends of vectors perpendicular to the lattice planes of a crystal and of length inversely proportional to their interplanar spacings.

Representative triangle: For any particular crystal system, the projection on the stereogram of the spherical triangle whose repetition by the maximum symmetry of the system would completely cover the sphere without overlap.

Screw axis: A symmetry element in a repeating pattern whose operation is a combination of a rotation about the axis by $360°/n$ and a shift parallel to the axis by m/n of the lattice translation, where $1 \leqslant m < n$.

Space group: A self-consistent combination of symmetry operations in a repeating pattern.

Sphere of reflection: A sphere centred at a point on the incident ray and passing through the origin of the reciprocal lattice, such that if a reciprocal lattice point lies on the sphere the corresponding set of lattice planes of the crystal satisfies the Bragg equation.

Spherical triangle: A triangular figure on the surface of a sphere bounded by three arcs of great circles.

Stereographic projection: A mapping of the surface of the sphere on to its equatorial plane by joining points on the sphere to the south pole.

Structure factor (F): The factor in the expression for the amplitude of a Bragg reflection that involves the coordinates of the atoms within one unit cell.

Symmetry element: A point, line or plane which is not moved by a symmetry operation, and with which the particular operation is associated.

Symmetry operation: An operation (such as rotation through an angle) after which an object is in a position indistinguishable from that which it occupied before.

Tetrad: A 4-fold rotation axis or rotation–inversion axis.

Triad: A 3-fold rotation axis or rotation–inversion axis.

Truncate: A small face equally inclined to two adjacent larger faces of a form is said to truncate the edge that would otherwise occur between the larger faces. Similarly a small face equally inclined to three or more faces of a form is said to truncate the vertex at which they would meet.

Twinned crystal: Composite crystal consisting of two or more differently oriented parts that are related to one another by a symmetry operation.

Unit cell: The geometrical block-like outline enclosing a unit of pattern in a crystal's structure.

Well-formed crystal: A crystal on which all faces of any particular form are of the same size.

White X-rays: X-rays having a wide range of wavelength.

Zone: A set of faces which are all parallel to a line and intersect in parallel edges.

Zone axis: The direction which is parallel to all the faces in a zone.

Zone symbol: Set of three whole numbers $[uvw]$ enclosed in square brackets, such that the coordinates of any point on the zone axis are in the ratio $ua:vb:wc$.

Answers to Problems

Chapter 1, p. 11
1. $a:b:c = 0.77:1:1.23$.
2. $F\hat{\ }B = 33°$, $C\hat{\ }S = 51°$.
3. $\alpha = 90°$, $\beta = 109°$, $\gamma = 90°$; $a:b:c = 0.80:1:1.29$.
4. $F\hat{\ }B = 33\frac{1}{2}°$, $C\hat{\ }S = 50\frac{1}{2}°$.

Chapter 2, p. 16
1. $a:b:c = 1.51:1:2.42$.
2. $a' = 2a$, $b' = b$, $c' = 2c$, i.e. the new cell is made up of four of the original cells. G is stepped $a' \times 2b'$ and S is stepped $2b' \times c'$. It would seem at first sight that they would be $\frac{1}{2}a' \times b'$ and $b' \times \frac{1}{2}c'$, but fractions of cells are not allowed so one must assume the steps to be twice as big.
3. $A\hat{\ }Q = 36°$.
4. $\alpha = 90°$, $\beta = 107°$, $\gamma = 90°$; $a:b:c = 0.79:1:1.27$.
5. The new cell has the same a- and b-edges as the old cell, but the new c-edge is the short diagonal of the ac face of the old cell. The face P is stepped $2a' \times c'$ in terms of the new cell.

Chapter 3, p. 28
1. Three 4-fold axes through opposite vertices; four 3-fold axes through the centres of opposite faces; six 2-fold axes through the mid-points of opposite edges; three mirror planes perpendicular to the 4-fold axes, and six more mirror planes perpendicular to the 2-fold axes; and a centre of symmetry (which can also be regarded as converting the 3-fold axes to $\bar{3}$-axes).
2. Four 3-fold axes through a vertex and the centre of an opposite face; three $\bar{4}$-axes through the mid-points of opposite edges; six mirror planes each containing an edge and the mid-point of the opposite edge.
3. A vertical 4-fold axis through the apex; two vertical mirror planes through opposite corners of the base and two vertical mirror planes through the mid-points of opposite sides of the base.
4. Three 2-fold axes through the centres of opposite faces; three mirror planes, each containing two of these 2-fold axes; and a centre of symmetry.
5. A 2-fold axis parallel to the shaft and two mirror planes at right-angles intersecting along the 2-fold axis.

Chapter 4, p. 35
1. (100), (010), (001), (101), $(10\bar{1})$, (110), (120), (011), $(1\bar{1}1)$.
2. $[001]$, $[\bar{1}01]$ (or $[10\bar{1}]$ which is the same thing), $[\bar{1}\bar{1}1]$; (121).
3. (100), (010), (001), (101), $(10\bar{1})$, (210), (110), (012), $(2\bar{1}2)$, $[001]$, $[\bar{1}01]$, $[1\bar{2}1]$; (111).
4. (101), (010), (001), (102), (100), (111), (121), (011), $(\bar{1}\bar{1}2)$; $[\bar{1}01]$, $[\bar{2}01]$, $[0\bar{1}1]$; (122).
5. (111), $(\bar{1}11)$, $(1\bar{1}1)$, $(11\bar{1})$, $(\bar{1}\bar{1}1)$, $(\bar{1}1\bar{1})$, $(1\bar{1}\bar{1})$, $(\bar{1}\bar{1}\bar{1})$, $\{111\}$, $[110]$, $[1\bar{1}0]$, $[101]$, $[10\bar{1}]$, $[011]$, $[01\bar{1}]$.

Chapter 5, p. 51
1. (Results to the nearest 100 km):
   London–Mexico 8900; –Delhi 6600; –Sydney 17,100;
   Mexico City–Delhi 14,600; –Sydney 13,000; Delhi–Sydney 10,400.
2. Cubic, with maximum (holosymmetric) symmetry. The faces all belong to two forms:
   a, c, e, g, j, m $\{100\}$
   u, v, s, t, w, x, y, z $\{111\}$.
3. 2468 km; W 37°1'S; W 39°10'S.
4. 56°28'; 46°41'; 1.5089:1:0.9430.

Chapter 6, p. 81
1. Triclinic (210), $(\bar{2}10)$. Monoclinic and orthorhombic (210), $(\bar{2}10)$, $(2\bar{1}0)$, $(\bar{2}\bar{1}0)$. Tetragonal (210), (120), $(\bar{2}10)$, $(\bar{1}20)$, $(2\bar{1}0)$, $(\bar{1}20)$, $(\bar{2}\bar{1}0)$, $(\bar{1}\bar{2}0)$. Cubic (210), (120), $(\bar{2}10)$, $(\bar{1}20)$, $(2\bar{1}0)$, $(1\bar{2}0)$, $(\bar{2}\bar{1}0)$, $(\bar{1}\bar{2}0)$, (201), (102), $(\bar{2}01)$, $(\bar{1}02)$, $(20\bar{1})$, $(\bar{1}0\bar{2})$, $(0\bar{2}1)$, (021), (012), $(0\bar{2}1)$, $(0\bar{1}2)$, $(02\bar{1})$, $(01\bar{2})$, $(0\bar{2}\bar{1})$, $(0\bar{1}\bar{2})$.
2. $(21\bar{3}0)$, $(12\bar{3}0)$, $(\bar{1}\bar{3}20)$, $(\bar{2}\bar{3}10)$, $(\bar{3}210)$, $(\bar{3}120)$, $(\bar{2}1\bar{3}0)$, $(\bar{1}2\bar{3}0)$, $(1\bar{3}20)$, $(2\bar{3}10)$, $(3\bar{2}10)$, $(3\bar{1}20)$.
3. $22°12'$.

Answers to Problems   247

4. Referred to hexagonal axes the faces are $(10\bar{1}2)$, $(\bar{1}102)$, $(0\bar{1}12)$, $(\bar{1}01\bar{2})$, $(1\bar{1}0\bar{2})$ and $(01\bar{1}\bar{2})$ which convert to (411), (141), (114), $(\bar{4}\bar{1}\bar{1})$, $(\bar{1}\bar{4}\bar{1})$, $(\bar{1}\bar{1}\bar{4})$.
5. $\beta = 99°18'$; $a:b:c = 0.6899 : 1 : 0.4124$.
6. If A is $\{100\}$ then B is $\{110\}$, C is $\{001\}$, D is $\{101\}$ and $c/a = 1.3270$.
   If B is $\{100\}$ then A is $\{110\}$, C is $\{001\}$, D is $\{111\}$ and $c/a = 0.9384$.

Chapter 7, p. 117
1. 222 and 2.
2. (i) There are no 3-fold axes along diagonals, and no 4-fold axes perpendicular to any face, so the crystal must be orthorhombic and may be $2/m\ 2/m\ 2/m$, $2mm$, or 222.
   (ii) $\bar{4}2m$ or $\bar{4}$.
   (iii) $\bar{4}$.
   (iv) The symmetry is incompatible with any system higher than monoclinic, and the right angles must be accidental and only approximate.
3. (i) $(112), (121), (211), (\bar{1}12), (\bar{1}21), (\bar{2}11), (1\bar{1}2), (1\bar{2}1), (2\bar{1}1), (11\bar{2}), (12\bar{1}), (21\bar{1})$ and repeat with the third index negative.
   (ii) $(112), (121), (211), (\bar{1}\bar{1}2), (\bar{1}\bar{2}\bar{1}), (\bar{2}\bar{1}\bar{1}), (\bar{1}1\bar{2}), (\bar{1}2\bar{1}), (\bar{2}1\bar{1}), (1\bar{1}\bar{2}), (1\bar{2}\bar{1}), (2\bar{1}\bar{1}) $.
   (iii) $(210), (\bar{2}10), (\bar{2}\bar{1}0), (2\bar{1}0), (021), (0\bar{2}1), (0\bar{2}\bar{1}), (02\bar{1}), (102), (\bar{1}0\bar{2}), (\bar{1}02), (10\bar{2})$.
   (iv) $(321), (213), (132), (\bar{3}2\bar{1}), (\bar{2}13), (\bar{1}32), (3\bar{2}1), (2\bar{1}3), (1\bar{3}2), (32\bar{1}), (21\bar{3}), (13\bar{2})$.
4. $3m$

Chapter 8, p. 129
1. 16.4 Å, 1.15 Å.
2. 13.9 Å, 0.97 Å.
3. 11.1 Å, 0.77 Å.
4. 5.1 Å, 0.36 Å.

Chapter 9, p. 144
1. 3.78 Å; 0.9037 Å.
2. 71%.

Chapter 10, p. 154
1. $F_{111} = 88.4$; $F_{331} = 46.0$.
2. (a) For 111 $M = 8$ $l \propto 1.457$ $L = 1.354$ $P = 0.764$.
       For 331 $M = 24$ $l \propto 1.462$ $L = 0.538$ $P = 0.766$.
   (b) For 111 $M = 8$ $l \propto 2.980$ $L = 2.937$ $P = 0.944$.
       For 331 $M = 24$ $l \propto 1.292$ $L = 1.166$ $P = 0.700$.
3. (a) 3.08. (b) 9.64.

Chapter 11, p. 166
1. F (all faces centred).
2. Indexing with B as $\{100\}$ would convert it to an I-centred cell of half the volume, which is more appropriate.
3. $a = 5.67$ Å, $c = 10.64$ Å.
4. A is $\{110\}$, B is $\{100\}$, C is $\{001\}$ and D is $\{112\}$. Thus the simple assumption that D was $\{111\}$ is shown to be inappropriate in terms of the true $c:a$ ratio determined by X-rays.

Chapter 12, p. 176
1. I-centred; $a = 9.41$ Å.
2. F-centred; $a = 8.40$ Å.
3. Apparently primitive with $a = 3.146$ Å. If the structure is similar to NaCl then the value of a should be about 6.24 Å, i.e. twice the value found. In this case all the indices found for the reflections should be doubled and they become:

$$200, 220, 222, 400, 420, 422, 440, 600, 620, 622, 444, 640, 642, 800.$$

The explanation is that K^+ and Cl^- each contain the same number of electrons and therefore have almost identical scattering factors (f). Thus the cell has virtually identical atoms (in this respect) at the corners, face centres and edge centres, and is almost indistinguishable from a stack of eight primitive cells of half the a-dimension so far as the X-rays are concerned.

Chapter 13, p. 193
1. $c = 6.18$ Å.
2. 100, 110, 200, 210, 220, 300, 310, 320, 400; $a = 5.36$ Å.
3. The indices derived in (2) correspond to $h^2 + k^2 = 1, 2, 4, 5, 8, 9, 10, 13, 16$. The reflection at 0.20 on the first-layer line shows that these values must all be doubled. Thus the zero layer line reflections are 110, 200, 220, 310, 400, 330, 420, 510 and 440; the first layer line reflections are then 101, 211, 301, 321, 411, 431 and 521; those on the second layer line are as for the zero layer line but with the third index 2, i.e. 112, etc.
   The Bravais lattice is I-centred, and $a = 7.58$ Å.
4. The y-axis. The X-ray beam makes angles between 0° and 14° with the x^*-axis of the reciprocal lattice within the acute angle β^*. It therefore makes angles between 9° and 23° with the x-axis of the crystal within the obtuse angle β.

Chapter 14, p. 205
1. $P2_1/c$.
2. $C2/c$.
3. P2 or Pm.

Chapter 15, p. 216
1. The z-coordinate of Mg is taken as zero. If the z-coordinate of oxygen is regarded as a variable, there are oxygen atoms at $\pm z$, and clearly $z \ll \frac{1}{2}$. Subject to this condition the most similar solutions for $F_{0002} = 0$ and $F_{0003} = 0$ are given by $z = 0.196$ and $z = 0.184$ respectively. With $z = 0.19$ both these structure factors are less than 1, and $F_{0001} > 15$. The required distance is therefore 1.8 Å.
2. Normal.

Chapter 16, p. 224
1. Zero layer line from centre to right: 020, 040, 060;
   Upper layer line (ditto): $31\bar{1}, 33\bar{1}, 35\bar{1}$;
   Lower layer line (ditto): $\bar{3}11, \bar{3}31, \bar{3}51$;
   The second index changes sign on the left-hand side.
2. The spots in the top right-hand quadrant are:

$$\begin{array}{ccc} 310 & 330 & 350 \\ 200 \quad 220 & 240 & 260 \\ 110 & 130 & 150 \\ 020 & 040 & 060 \end{array}$$

   The first index changes sign in the lower half of the photograph and the second index changes sign on the left-hand side.
3. C-centred.
4. About 25°. Indices of spots in the upper right-hand quadrant are

$$\begin{array}{ccc} 31\bar{3} & 33\bar{3} & 35\bar{3} \\ 20\bar{2} \quad 22\bar{2} & 24\bar{2} & 26\bar{2} \\ 11\bar{1} & 13\bar{1} & 15\bar{1} \\ 020 & 040 & 060 \end{array}$$

   The first and third indices change sign in the lower half of the photograph and the second index changes sign on the left-hand side.
5. $20\bar{3}$. The beam is parallel to the planes $(20\bar{3})$ and (020) and therefore to the zone axis $[302]$.

Further Reading

For further reading beyond the scope of this book the reader is referred to the following works on the specific topics indicated.

Symmetry in crystals:
  Elementary Crystallography. M. J. Buerger. M.I.T. Press, 1978.

Illustrations of combinations of forms in each crystal class:
  Dana's Textbook of Mineralogy (4th ed.) W. E. Ford. Wiley, 1963.

Measurement of crystal morphology and projection methods:
  Crystallometry. P. Terpstra and L. W. Codd. Longmans, 1960.

X-ray crystallography (general):
  X-Ray Diffraction Methods. E. W. Nuffield. Wiley, 1966.
  X-Ray Crystallography. G. H. W. Milburn. Butterworths, 1973.
  An Introduction to Crystallography. M. M. Woolfson. Cambridge University Press, 1970.

X-ray powder methods:
  The Powder Method in X-Ray Crystallography, L. V. Azaroff and M. J. Buerger. McGraw Hill, 1958.

X-ray single crystal photographic methods:
  X-Ray Crystallography. M. J. Buerger. Wiley, 1942.
  Contemporary Crystallography. M. J. Buerger. McGraw-Hill, 1970.
  The Precession Method. M. J. Buerger. Wiley, 1964.

Single crystal diffractometry for both X-rays and neutrons:
  Single Crystal Diffractometry. U. W. Arndt and B. T. M. Willis. Cambridge University Press, 1966.

Electron diffraction:
  Interpretation of Electron Diffraction Patterns (2nd Ed.). K. W. Andrews, D. J. Dyson and S. R. Keown. Hilger, 1971.

The determination of crystal structures:
  X-Ray Structure Determination. G. H. Stout and L. H. Jenson. Macmillan, 1968.

The results of crystal structure determination, with close reference to the crystallographic features:
  Crystal Structures: A Working Approach. H. D. Megaw. W. B. Saunders Co., 1973.
  Crystal Structure of Minerals. W. L. Bragg and C. F. Claringbull. Bell, 1965.
  Structural Inorganic Chemistry (3rd Ed.). A. F. Wells. Oxford University Press, 1962.

The history of X-ray crystallography:
  The Development of X-Ray Analysis. W. L. Bragg. Bell, 1975.

Index

Absent reflections
  for non-primitive cells   163
  for hexagonal lattice   166
  for rhombohedral lattice   166
Absorption factor   153
Amplitude
  of a wave   145
  of reflection   147
Anorthic system   20
Arc length factor   150
Atomic scattering factor   145, 244
Axes
  crystallographic   29
  of a crystal   244
Axial ratios   8
Axis
  of rotation   17, 244
  of rotation–inversion   19, 115, 244

Back-reflection camera   133
Basal pinacoid   57
Beam-trap   130
Bernal chart   188, 189
Bevel   244
Bipyramid   54
  trigonal   107, 108
Bisphenoid   89
Body-centred cell   156
Body-centred lattice   158
  absent reflections from   164
Boundaries, low angle   237
Bragg angle   123, 245
Bragg equation   122, 245
Bravais lattices   155, 162
Bravais, Law of   241
Broad reflections   143

C-centred cell   156
C-centred lattice   155
  absent reflections from   164
Centre of symmetry   19, 244
Centred cell   244
Centred lattice   156
  symmetry in   162
Centrosymmetric classes   86
Characteristic symmetry of a crystal system   24
Chrysotile, disorder in   234
Circle of reflection   185, 244

Class, determination of   110
Class 1   86
Class $\bar{1}$ 86
Class 2   87
Class m 87
Class $2/m$ 87
Class 222   89
Class $mm2$ 89
Class $2/m\ 2/m\ 2/m$ 89
Class mmm 89
Class 4   93
Class 422   92
Class $4mm$ 92
Class $4/m$ 93
Class $\bar{4}$ 95
Class $\bar{4}2m$ 95
Class $4/m\ 2/m\ 2/m\ (4/mmm)$ 92
Class 23   96
Class 432   96
Class 43   98
Class $2/m\ \bar{3}$ 96
Class $\bar{4}3m$ 96
Class $4/m\ \bar{3}\ 2/m$ 96
Class $m3m$ 96
Class 6   100
Class 622   100
Class $6mm$ 100
Class $6/m$ 100
Class $6/m\ 2/m\ 2/m$ 100
Class 3   107
Class 32   107
Class $3m$ 107
Class $\bar{3}$ 109
Class $\bar{3}\ 2/m$ 109
Clay minerals, disorder in   234
Cleavage   16, 225
Collimator   130
Composition plane   223
Compositional disorder   235
Contact twin   226
Constancy of angles   245
Crystal class   26, 82, 244
  nomenclature of   84
Crystal growth   83, 241
Crystal system   23, 244, 245
Crystal, well-formed   36
Crystallographic axes   29
Crystallographic symmetry elements   19
Cube form   60, 101
Cubic classes   96

Index

Cubic holosymmetric class, forms in 64
Cubic powder pattern
  for centred lattices 171
  indexing of 169
Cubic system 22

Dauphiné twinning 230
Debye–Scherrer camera 130
Dendritic growth 242
Diad 25, 244
Diffractometer, powder 136
Dihexagonal bipyramid 75, 103
Dihexagonal prism 75, 78, 103, 109
Dihexagonal pyramid 103
Direction cosines 66, 244
Dislocations 236
Disphenoid 89
Ditetragonal bipyramid 56, 94
Ditetragonal prism 57, 94, 105
Ditetragonal pyramid 56
Ditetragonal sphenoid 95
Ditrigonal pyramid 107, 108
Ditrigonal scalenohedron 80, 95, 109
Dodecahedron
  pentagonal 99, 101
  rhombic 61, 101
Donnay–Harker Law 241

Edge 3, 244
Edge dislocation 236
Electron diffraction 217
  patterns, indexing of 220
  specimens for 219
Electron micrographs, high resolution 223
Electron microscope 218
Electron wavelengths 217
Element of symmetry 17
Elements of crystallographic symmetry 19
Enantiomorphous 244
Enantiomorphous classes 86
Etch pits 111

F-centred lattice 159
  absent reflections from 165
Face 3, 245
  corresponding 3
Face-centred lattice 159
  absent reflections from 165
Face indices 30
Filtered radiation 128
Form 30, 244
  closed 31
  indices 31, 244
  open 31
  and growth conditions 243
  and structure 240

Gandolfi camera 134

General form 54
Glide plane 196, 244
Growth of crystals 241
Growth twinning 228
Guinier camera 134

Habit 244
Hexad 26, 244
Hexagonal bipyramid 75, 78, 103, 105, 109
Hexagonal classes 100
Hexagonal holosymmetric class, forms in 75, 103
Hexagonal indexing
  conversion to rhombohedral indexing 78
  in trigonal system 81
Hexagonal lattice parameters, conversion to rhombohedral 166
Hexagonal prism 75, 78, 103, 105, 107, 108, 109
Hexagonal pyramid 103, 107, 108
Hexagonal system 22, 23
Hexagonal trapezohedron 103
Holosymmetric 244
Holosymmetric class 53
Hopper crystals 242

I-centred cell 156
I-centred lattice 158
  absent reflections from 164
Ill-formed crystal 245
Indices
  of a face 30, 244
  Miller 30
  Miller–Bravais 74
Interfacial angles 10, 244
  of the cubic system 66
Intergrowth, oriented 242
Interpenetrant twin 226, 244
Interplanar spacing 123
Interstitials 236
Inversion 19

$K\alpha$ wavelengths 127

Laün groups 115
Law of Bravais 241
Lattice 245
  definition of 14, 83
Lattice plane 24, 121
Layer lines 181
Lorentz factor 152
  derivation of 192
Low angle boundaries 237
Low symmetry classes, forms in 86

Miller–Bravais indices 74
Miller indices 30
  ($hkil$) in the hexagonal system 74

Minimum symmetry of a crystal system 23
Mirror line 17
Mirror plane 17, 245
Monochromatic X-rays 121, 245
Monochromator 135
Monoclinic classes 87
   holosymmetric class, forms in 67, 88
Monoclinic system 20
   calculations in 68
   stereogram of 67
Mosaic structure 238
Multiplicity factor 150

Napier's rule 50
Neutron diffraction 214
Neutron scattering factors 214
Non-centrosymmetric classes 86
Nucleation twinning 230

Occupancy refinement 214
Octahedron form 60, 101
Optical activity 111
Ordering in solid solutions 214
Orthorhombic classes 89
Orthorhombic holosymmetric class, forms in 56, 91
Orthorhombic system 22
Oscillation photograph 184
   indexing of 186

P-lattice 156
Parallelepiped 20
Particle size broadening 143
Particle size determination 143
Parting 225
Pedion 87
Pentagonal dodecahedron 99, 101
Piezo electricity 114
Pinacoid 31, 245
Point group 245
   nomenclature of 84
Polar 245
Polarization factor 151
Pole of a face 37
Polysynthetic twinning 229
Polytypes 232
Powder camera 130
Powder diffractometer 136
Powder pattern
   cubic 169
   non-cubic 173
Powder photograph 124
Preferred orientation 133, 143
Primitive cell 245
Primitive lattice 156
Primitive of the stereographic projection 38, 245
Prism 245
   ditrigonal 105
   form 55
   trigonal 107, 108

Pyritohedron 99
Pyramid
   ditrigonal 107, 108
   trigonal 107, 108
Pyro-electricity 114

Qualitative analysis by X-ray diffraction 139

Reciprocal lattice 182, 245
Reflection
   circle of 185
   sphere of 187
Repeating pattern, choice of cell in 83
Repeating patterns 12
Representative triangle 53, 245
Residual (R) 206
Reticular density 241
Rhombic-dodecahedron 61, 101
Rhombohedral indexing
   conversion to hexagonal indexing 81
   in trigonal system 78
Rhombohedral lattice parameters, conversion to hexagonal 166
Rhombohedron 22, 79, 80, 107, 108, 109
Rotation axis 17
Rotation–inversion, axis of 19, 115, 244
Rotation–inversion axes, properties of 115
Rotation photograph 179
   indexing of 182
Rotation point 17
Rutile, structure determination of 207

Scattering factor, atomic 145
Screw axis 194, 245
Screw dislocations 237
   and crystal growth 243
Shear twinning 229
Sketch stereograms 53
Solid solutions
   determination of 141
   structure refinement of 213
Soller slits 138
Space group 197, 245
   absences 200
   determination 198
   representation 200
   symbols 197
Special form 55
Sphenoid 89
Sphere of reflection 187, 245
Spherical projection 37
Spherical triangle 245
   right angled 50
   solution of 48
Spherical trigonometry 48
Stacking disorder 232
   diffraction from 232
Stereogram 44

Stereographic net   42
Stereographic projection   38, 245
Stereoscopic drawings, viewing of   27
Striations   110
Structure factor   149, 245
Structure refinement   206
Symmetry element   246
  symbols of   19, 197
Symmetry operation   17, 245
Systems, characteristic symmetry of   24

Temperature factor   150
Tetrad   26, 245
Tetragonal bipyramid   57, 94
Tetragonal bisphenoid   95
Tetragonal classes   92
Tetragonal holosymmetric class, forms in   57, 94
Tetragonal prism   57, 94
Tetragonal sphenoid   95
Tetragonal system   22
Tetragonal trapezohedron   93
Tetrahedron form   99, 101
Tetrahexahedron   64, 99, 101
Transformation twinning   230
Translation symmetry   194
Transmission camera   133
Trapezohedron   93
  hexagonal   103
  trigonal   107, 108
Triad   26, 245
Triclinic classes   86
Triclinic system   20
  stereogram of   71
Trigonal bipyramid   105, 107, 108
Trigonal classes   107
Trigonal prism   105
Trigonal pyramid   107, 108
Trigonal system   23
Trigonal trapezohedron   107, 108
Trigonometry, spherical   48
Trisoctahedron   64, 101
Truncate   245

Turbostratic stacking   233
Twin axis   228
Twin law   228
Twin plane   228
Twinned crystal   245
Twinning   226

Unit cell   245
  alternative choices   15
  contents, symmetry of   83
  definition of   14
  dimension, determination of in cubic   172
Unit cells
  three-dimensional symmetry of   20
  two-dimensional symmetry of   19

Vacancies   236

Weber symbols for zone axes in the hexagonal system   74
Well-formed crystal   36, 245
White X-rays   121, 245
Wulff net   43

X-ray absorption   127
X-ray Powder Data File   139
X-ray scattering   121
X-ray spectrum   126
X-ray tube   125
X-ray wavelength   126
X-rays
  generation of   125
  monochromatic   121
  white   121
ξ(xi)-coordinate   189

ζ(zeta)-coordinate   189
Zone   3, 245
Zone axis   3, 32, 245
Zone symbol   24, 33